SCHOLASTIC

English & Maths Bumper Book Ages 9–10

Revision & Practice

Master maths and English topics with ease

First published in the UK by Scholastic, 2016; this edition published 2023

Scholastic Distribution Centre, Bosworth Avenue, Tournament Fields, Warwick, CV34 6UQ

Scholastic Ireland, 89E Lagan Road, Dublin Industrial Estate, Glasnevin, Dublin, D11 HP5F

www.scholastic.co.uk

1 2 3 4 5 6 7 8 9   3 4 5 6 7 8 9 0 1 2

A CIP catalogue record for this book is available from the British Library.

ISBN 978-0702-32677-6
Printed and bound by Bell and Bain Ltd, Glasgow

This book is made of materials from well-managed, FSC-certified forests and other controlled sources.

Authors

Lesley and Graham Fletcher (English) and Paul Hollin (Maths)

Editorial team

Rachel Morgan, Vicki Yates, Audrey Stokes, Tracey Cowell, Rebecca Rothwell, Jane Jackson, Sally Rigg, Jenny Wilcox, Mark Walker, Red Door Media Ltd, Kate Baxter, Margaret Eaton and Julia Roberts

Design team

Dipa Mistry, Andrea Lewis, Nicolle Thomas, Neil Salt and Oxford Designers and Illustrators

Illustration
Judy Brown and Simon Walmesley

Contents

How to use

This book has been written to help children reinforce the English and maths skills they have learned in school. Each subject is divided into sections covering a range of topics from the National Curriculum. Use the book little and often to practise skills and increase confidence. You can choose to work through the English and maths sections in order or focus on specific topics.

At the back of the book is a **Progress tracker** to enable you to record what has been practised and achieved.

English

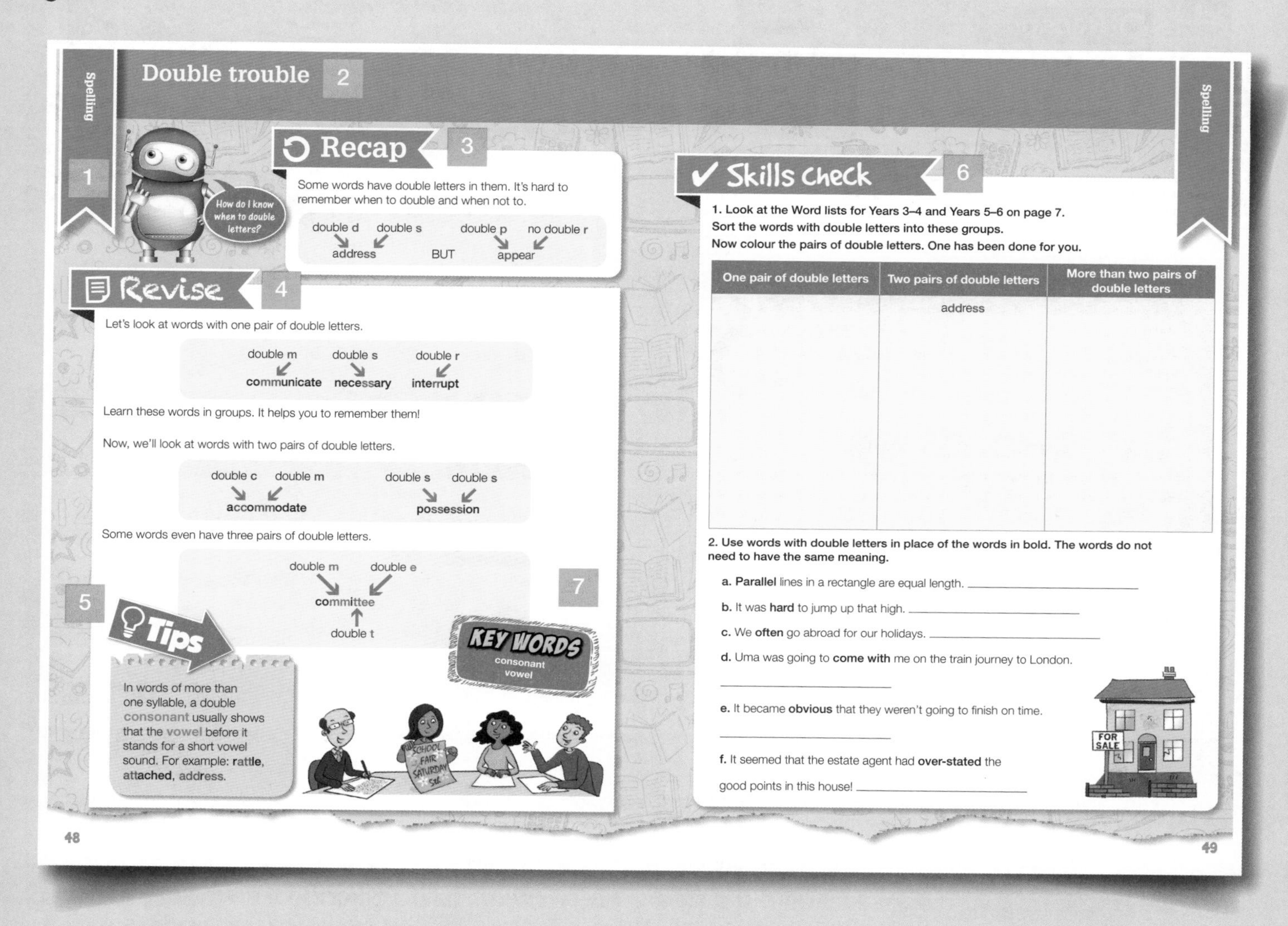

Spelling

Double trouble 2

1

How do I know when to double letters?

Recap 3

Some words have double letters in them. It's hard to remember when to double and when not to.

double d, double s → address BUT double p, no double r → appear

Revise 4

Let's look at words with one pair of double letters.

double m → communicate; double s → necessary; double r → interrupt

Learn these words in groups. It helps you to remember them!

Now, we'll look at words with two pairs of double letters.

double c, double m → accommodate; double s, double s → possession

Some words even have three pairs of double letters.

double m, double e, double t → committee

5

Tips

In words of more than one syllable, a double **consonant** usually shows that the **vowel** before it stands for a short vowel sound. For example: **rattle**, **attached**, **address**.

7

KEY WORDS

consonant
vowel

SCHOOL FAIR SATURDAY

48

Spelling

Skills check 6

1. Look at the Word lists for Years 3–4 and Years 5–6 on page 7. Sort the words with double letters into these groups. Now colour the pairs of double letters. One has been done for you.

One pair of double letters	Two pairs of double letters	More than two pairs of double letters
	address	

2. Use words with double letters in place of the words in bold. The words do not need to have the same meaning.

a. **Parallel** lines in a rectangle are equal length. __________

b. It was **hard** to jump up that high. __________

c. We **often** go abroad for our holidays. __________

d. Uma was going to **come with** me on the train journey to London. __________

e. It became **obvious** that they weren't going to finish on time. __________

f. It seemed that the estate agent had **over-stated** the good points in this house! __________

FOR SALE

49

1 Chapter title

2 Topic title

3 Each page starts a **recap** with a 'What is...' question which gives children a clear definition for the terminology used.

4 In the **revise** section there are clear explanations and examples, using clear illustrations and diagrams, where relevant.

5 **Tips** provide short and simple advice to aid understanding.

6 The **skills check** sections enable children to practise what they have learned with answers at the back of the book.

7 **Key words** that children need to know are displayed. Definitions for these words can be found in the **Glossary**.

Maths

The Maths section has many of the same features of the English section and also some additional ones. Keep some blank or squared paper handy for notes and calculations!

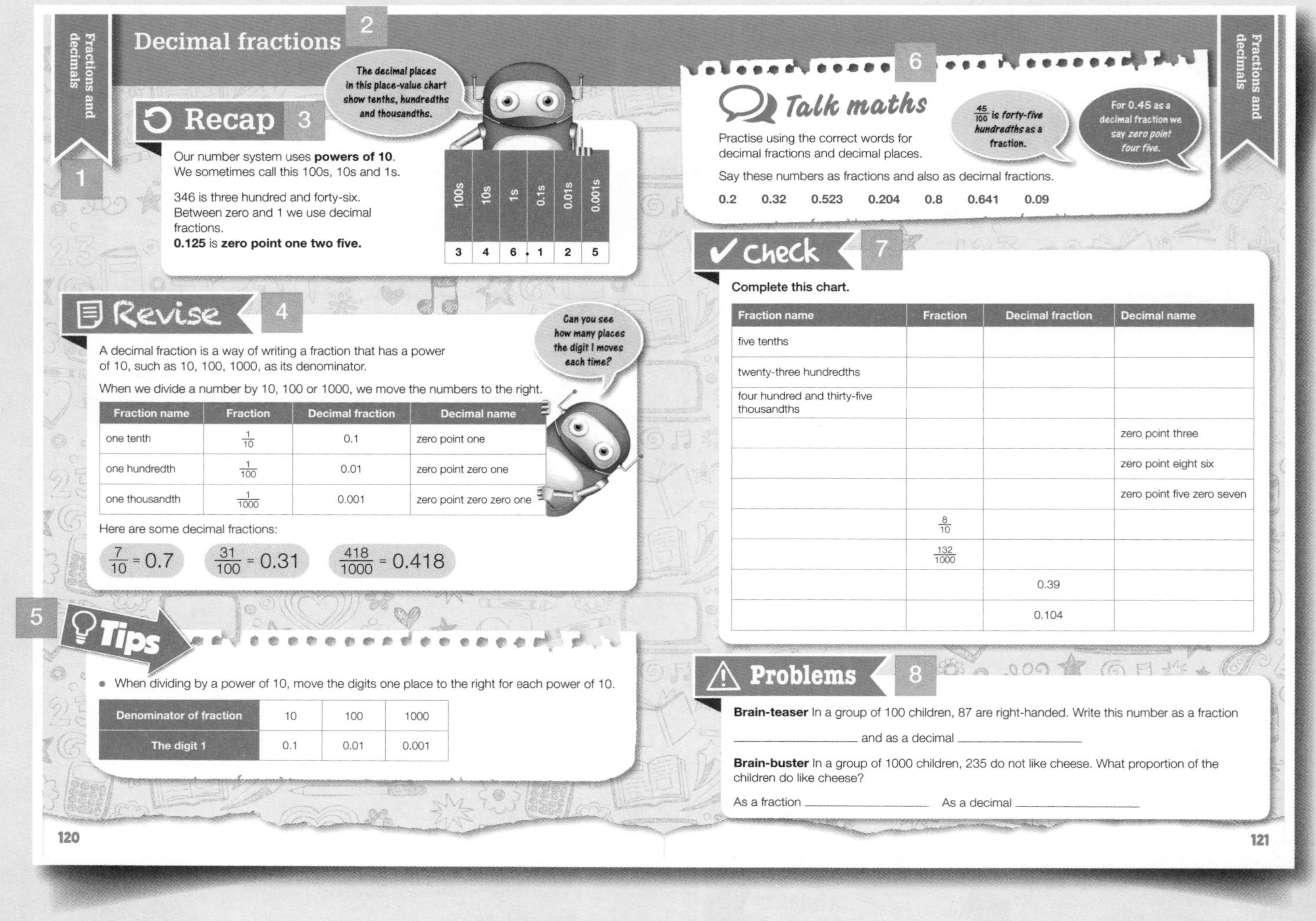

Fractions and decimals

1

Decimal fractions

2

Recap

3

Our number system uses **powers of 10**. We sometimes call this 100s, 10s and 1s.

346 is three hundred and forty-six. Between zero and 1 we use decimal fractions.

0.125 is **zero point one two five.**

The decimal places in this place-value chart show tenths, hundredths and thousandths.

100s	10s	1s	0.1s	0.01s	0.001s
3	4	6 .	1	2	5

Revise

4

A decimal fraction is a way of writing a fraction that has a power of 10, such as 10, 100, 1000, as its denominator.

When we divide a number by 10, 100 or 1000, we move the numbers to the right.

Can you see how many places the digit 1 moves each time?

Fraction name	Fraction	Decimal fraction	Decimal name
one tenth	$\frac{1}{10}$	0.1	zero point one
one hundredth	$\frac{1}{100}$	0.01	zero point zero one
one thousandth	$\frac{1}{1000}$	0.001	zero point zero zero one

Here are some decimal fractions:

$\frac{7}{10} = 0.7$ $\frac{31}{100} = 0.31$ $\frac{418}{1000} = 0.418$

5

Tips

- When dividing by a power of 10, move the digits one place to the right for each power of 10.

Denominator of fraction	10	100	1000
The digit 1	0.1	0.01	0.001

120

Fractions and decimals

6

Talk maths

Practise using the correct words for decimal fractions and decimal places.

$\frac{45}{100}$ is forty-five hundredths as a fraction.

For 0.45 as a decimal fraction we say zero point four five.

Say these numbers as fractions and also as decimal fractions.

0.2 0.32 0.523 0.204 0.8 0.641 0.09

Check

7

Complete this chart.

Fraction name	Fraction	Decimal fraction	Decimal name
five tenths			
twenty-three hundredths			
four hundred and thirty-five thousandths			
			zero point three
			zero point eight six
			zero point five zero seven
	$\frac{8}{10}$		
	$\frac{132}{1000}$		
		0.39	
		0.104	

Problems

8

Brain-teaser In a group of 100 children, 87 are right-handed. Write this number as a fraction ______________ and as a decimal ______________

Brain-buster In a group of 1000 children, 235 do not like cheese. What proportion of the children do like cheese?

As a fraction ______________ As a decimal ______________

121

1 Chapter title

2 Topic title

3 Each page starts a **recap** of basic facts of the mathematical area in focus.

4 In the **revise** section there are clear explanations and examples, using clear illustrations and diagrams, where relevant.

5 **Tips** provide short and simple advice to aid understanding.

6 **Talk maths** are focused activities that encourage verbal practice.

7 **Check** a focused range of questions, with answers at the end of the book.

8 **Problems** word problems requiring mathematics to be used in context.

Tips for using this book at home

Using this book, alongside the maths and English being done at school, can boost children's mastery of the concepts. Be sure not to get ahead of schoolwork or to confuse your child.

Keep sessions to an absolute maximum of 30 minutes. Even if children want to keep going, short amounts of focused study on a regular basis will help to sustain learning and enthusiasm in the long run.

Word lists These are the words you need to learn to spell.

Years 3–4

accident
accidentally
actual
actually
address
answer
appear
arrive
believe
bicycle
breath
breathe
build
busy/business
calendar
caught
centre
century

certain
circle
complete
consider
continue
decide
describe
different
difficult
disappear
early
earth
eight/eighth
enough
exercise
experience
experiment
extreme

famous
favourite
February
forward/
forwards
fruit
grammar
group
guard
guide
heard
heart
height
history
imagine
important
increase
interest

island
knowledge
learn
length
library
material
medicine
mention
minute
natural
naughty
notice
occasion
occasionally
often
opposite
ordinary
particular

peculiar
perhaps
popular
position
possess
possession
possible
potatoes
pressure
probably
promise
purpose
quarter
question
recent
regular
reign
remember

sentence
separate
special
straight
strange
strength
suppose
surprise
therefore
though/
though
thought
through
various
weight
woman/
women

Years 5–6

accommodate
accompany
according
achieve
aggressive
amateur
ancient
apparent
appreciate
attached
available
average
awkward
bargain
bruise
category
cemetery
committee

communicate
community
competition
conscience
conscious
controversy
convenience
correspond
criticise
curiosity
definite
desperate
determined
develop
dictionary
disastrous
embarrass
environment

equip
equipment
equipped
especially
exaggerate
excellent
existence
explanation
familiar
foreign
forty
frequently
government
guarantee
harass
hindrance
identity
immediate

immediately
individual
interfere
interrupt
language
leisure
lightning
marvellous
mischievous
muscle
necessary
neighbour
nuisance
occupy
occur
opportunity
parliament
persuade

physical
prejudice
privilege
profession
programme
pronunciation
queue
recognise
recommend
relevant
restaurant
rhyme
rhythm
sacrifice
secretary
shoulder
signature
sincere

sincerely
soldier
stomach
sufficient
suggest
symbol
system
temperature
thorough
twelfth
variety
vegetable
vehicle
yacht

Multiplication table

x	1	2	3	4	5	6	7	8	9	10	11	12
1	1	2	3	4	5	6	7	8	9	10	11	12
2	2	4	6	8	10	12	14	16	18	20	22	24
3	3	6	9	12	15	18	21	24	27	30	33	36
4	4	8	12	16	20	24	28	32	36	40	44	48
5	5	10	15	20	25	30	35	40	45	50	55	60
6	6	12	18	24	30	36	42	48	54	60	66	72
7	7	14	21	28	35	42	49	56	63	70	77	84
8	8	16	24	32	40	48	56	64	72	80	88	96
9	9	18	27	36	45	54	63	72	81	90	99	108
10	10	20	30	40	50	60	70	80	90	100	110	120
11	11	22	33	44	55	66	77	88	99	110	121	132
12	12	24	36	48	60	72	84	96	108	120	132	144

English Made Simple Ages 9–10

Adjectives

What is an adjective?

Recap

An **adjective** describes a characteristic of a noun.

the **blue** hat

The word **blue** is an adjective.

Revise

Adjectives describe or modify nouns. They give us more detail.

the **enormous** elephant

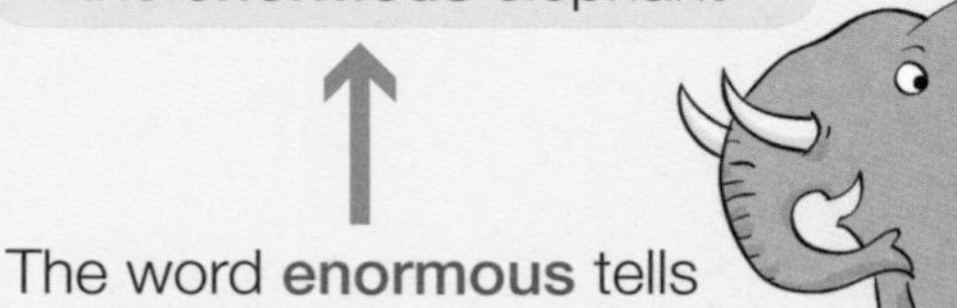

The word **enormous** tells us more about the elephant.

the **blue** elephant

The word **blue** tells us about a very different elephant!

It was an **unbelievable** story!

It was an **enthralling** story!

Not all adjectives describe characteristics we can see.

Adjectives often come before a noun. You can have more than one adjective to describe a noun.

✔ Skills Check

1. **Underline the adjectives.**

   a. They could hear the plane's supersonic engine.

   b. He filled in the necessary paperwork, before applying for a passport.

   c. She had a look of intense concentration on her face.

2. **Replace the word 'good' in the sentences on the right with a more interesting adjective. Write the new sentence.**

   a. It was a very **good** meal.

   Yummy

   b. Nikita had **good** results in her tests.

   Amazing

   c. Their family had a **good** holiday in Majorca.

   Fun

Nouns

Recap

A **noun** is a word for a person, place or thing. A noun is a naming word.

Revise

There are different types of noun: **common nouns** and **proper nouns**.

Common nouns

cat

house

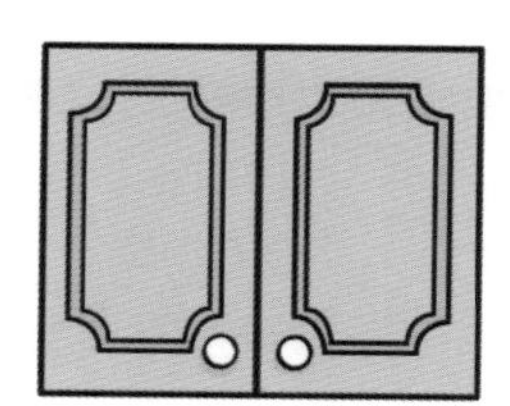
cupboard

Proper nouns

Names of people		
Jack	Sunita	Mrs Grey

Names of places		
London	France	River Tyne

Names of days of the week and months		
Wednesday	Monday	January

Tips

1. All proper nouns must start with a capital letter.
2. **Adjectives** + **nouns** make a **noun phrase**:

   quiet Sunday
   young William
   cheerful Mrs Grey

   You can use more than one adjective:

   a **hot windy** day
   a **large aggressive** dog
   a **long arduous** task

✓ Skills Check

1. Underline the nouns in these sentences.

a. Ellie cautiously opened the dark cupboard.

b. The American group finally reached the top of Mount Everest.

2. Put two adjectives in front of each noun to make noun phrases.

a. Beautiful, sunny afternoon

b. Deep, dark river

c. Green, scary crocodile

Verbs: tenses

Recap

What is a verb?

A **verb** tells you what is happening in a sentence. It is a doing word or being word.

The **tense** of a verb tells us when it happens: in the **present**, the **past** or the **future**.

What is a tense?

Revise

Action verbs → I run I read

Simple present tense
I run
he runs
we run

Simple past tense
I ran
she ran
we ran

Use simple present or past for an action happening now (present) or an action that has already happened (past).

Being verbs → I am I have

Present progressive tense
I **am** running

Past progressive tense
I **was** running

Use a **helper verb** (**to be** or **to have**) to show the action is/was continuous.

Present perfect
has/have + verb
He **has read** a book.
↑
action more recently in the past

Past perfect
had + verb
He **had read** the book.
↑
action further in the past

Skills Check

KEY WORDS
verbs
tense (past, present)
progressive
future
perfect

1. Underline the verbs in this sentence.

The dog was barking loudly when the postman brought a letter.

2. Fill in the missing verbs in the table below.

Present tense	Past tense	Present progressive
he eats	she ate	he is eating
she runs	they slept	she is sleeping
they look	they ran	we are running

3. Complete the sentence using the verb 'to finish' in the past perfect form.

They were finishing their tea when the phone rang.

Verbs: tense consistency and Standard English

Recap

What is tense consistency?

Tense consistency means having the same tense within a sentence.

What is Standard English?

Standard English is when the verb ending agrees with the thing or person doing the action. Standard English does not use slang or dialect words.

Revise

Tense consistency

Use only one tense in a sentence:

Ahmed **won** the race and everyone **applauded**. verbs – both in past tense

Standard English

A **singular** subject (or person doing it) must have a singular form of the verb:

Jack **reads** a book. → Jack **was reading** a book.
one person = singular form of verb

A **plural** subject (or things doing it) must have a plural form of the verb:

The children **read** a book. → The children **were reading** a book.
many people = plural form of verb

Make the verb ending agree with the number of doers!

✓ Skills Check

1. Rewrite the sentence so that the tenses are consistent.

Omar and Ranvir eats their lunch and goes out to play.

Omar and Ranvir were eating their lunch ang went to play.

2. Circle the Standard English form of the underlined verbs.

a. I **were** / **was** preparing a spicy curry for our tea.

b. George and Ahmed **is** / **are** outstanding cricketers.

c. Despite attractive adverts the property **appear** / **appears** dilapidated.

Modal verbs

Recap

Modal verbs go before other verbs. The modal verbs are:

can	could	would
shall	may	might
should	must	will

A modal verb expresses degrees of possibility: I **could** play football.

Revise

Modal verbs tell us how likely an action is.

1. Whether someone is able to do something:

Ellie **can** read in assembly.

2. How likely something is:

He **must** walk the dog, after tea.

They express degrees of certainty.
Must is more certain than **could**. **May** is less certain than **will**.

✓ Skills Check

KEY WORDS
modal verbs

1. Choose the best modal verb to complete each sentence.

a. Can ____________ I go to the bathroom, please?

b. We should ____________ go to the cinema this afternoon.

c. They ~~are~~ will ____________ be going on holiday on Saturday.

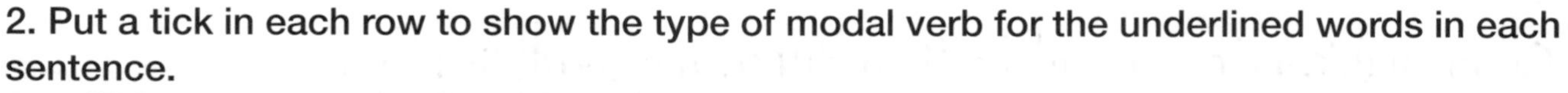

2. Put a tick in each row to show the type of modal verb for the underlined words in each sentence.

Sentence	Modal verb of possibility	Modal verb of certainty
Sunita <u>should</u> tidy her bedroom.	should✓	
Sunita <u>must</u> tidy her bedroom.		must✓
Sunita <u>might</u> tidy her bedroom.	might✓	
Sunita <u>can</u> tidy her bedroom.		can✓

Adverbs

What is an adverb?

Recap

An **adverb** describes a verb. It tells us *how* something was done.

Revise

Adverbs give us more detail about how a verb was done.
Adverbs often go next to the verb, but may go somewhere else in the sentence.

Jane ran. The sentence tells us she ran, but not *how* she ran.

Jane ran **slowly**. 

Jane ran **swiftly**.

Jane ran **cautiously**

These **adverbs** describe the verb **ran**. Adverbs often end in **ly**. Each adverb changes how Jane ran.

Adverbs describe the verb. They often end in ly.

✔ Skills Check

1. Underline the adverb in each sentence.

a. The waves lapped gently around her feet.

b. The music blared deafeningly from the large speakers.

c. Aggressively, the dog guarded his territory.

2. Choose an appropriate adverb to fit in each space.

a. The snow fell gently during the afternoon.

b. The sleeping baby snuffled quietly.

c. The footballer amazingly tackled the opposing team.

Adverbs and adverbials

What is an adverbial phrase?

Recap

An **adverbial** phrase tells us *how, where* or *when* something happened.

Revise

Be careful! Some words may be prepositions, such as back or up. Check how the word is used.

An adverbial phrase tells us:

- *how* it was done (manner) – **She walked** with great enthusiasm.
- *where* it was done (place) – **She walked** through the forest.
- *when* it was done (time) – After lunch **she went for a walk.**

Here are some examples of different types of adverb.

Adverbs of time	Adverbs of place	Adverbs of manner
soon	about	some adverbs ending **ly**, such as happily, friendly, greedily
before	indoors	fast
already	outside	hard
finally	anywhere	so
eventually	where	straight
next	towards	well
tomorrow	upstairs	
yesterday	near	
since	far	

KEY WORDS
adverbials

Skills Check

1. Put a circle round each adverbial phrase.

a. During the morning she received a telephone call.

b. There was a fire alarm in the shopping centre.

c. The cat curled up very gracefully.

2. Put a tick in each row to show the type of adverbial within each sentence.

Sentence	Adverbial of time	Adverbial of place	Adverbial of manner
Thomas played the piano really beautifully.			✓
This afternoon we will go to the park.		✓	
The traffic flowed over the bridge.	✓		

Tips

Does a word or phrase describe *how, when* or *where* something was done? If it does, it's an adverb or adverbial phrase.

Adverbs of possibility

Recap

What is an adverb of possibility?

An adverb of possibility shows how certain we are about something.

The most common adverbs of possibility are:

probably | perhaps | maybe | certainly
definitely | obviously | clearly | possibly

Revise

Maybe and **perhaps** usually come at the beginning of a sentence or clause.

Perhaps our visitors will arrive soon.

Maybe we can go for a walk, if it stops raining.

Other adverbs of possibility usually come in front of the main verb.

They **probably** will be going to France this summer.

However, they come after the verbs **am**, **is**, **are**, **was** and **were**.

We are **definitely** going to the party.

Skills Check

1. Choose an appropriate adverb of possibility for each sentence.

obviously | maybe | probably

a. Opposite sides of a rectangle are obviously ______ equal lengths.

b. Mabye ______ the water will be warm enough to swim in.

c. There is probably ______ enough petrol in the car.

2. Write a sentence, using an adverb of possibility.

____

Fronted adverbials

What is a fronted adverbial?

Recap

A **fronted adverbial** is an adverb or an adverbial phrase which is at the beginning of a sentence.

Revise

Before tea, let's go for a walk.

Adverbial phrase at the beginning of the sentence. **Adverbial of time** describes *when* we will go.

Rest of sentence.

In the middle of the square, they found a restaurant.

Adverbial of place describes *where* they found it.

Fortunately, there was no more snow.

Adverb of manner, also a fronted adverbial.

Rest of sentence.

A fronted adverbial comes at the front of the sentence.
An adverbial phrase can come anywhere in the sentence.

Skills Check

KEY WORD
fronted adverbials

1. Underline the fronted adverbial in this sentence.

During the winter, the geese arrived on the coastal marshes.

2. Write a fronted adverbial to complete the sentence below.

________________________________ the winds blew the trees down.

Main and subordinate clauses

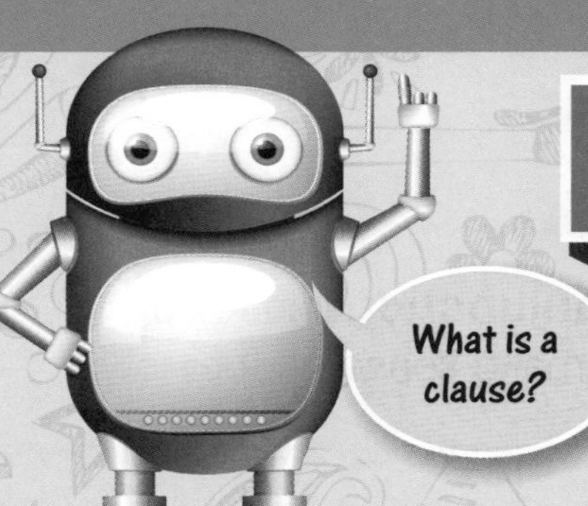

What is a clause?

Recap

A **clause** is a group of words which contain a subject (a person or thing who does the verb) and verb.

Revise

There are different types of clause.

- A **main clause** is an independent clause which makes sense by itself.
- A **subordinate clause** is dependent on the main clause to make sense.

The tent blew down.

A **main clause** can be a complete sentence. It has a subject (**the tent**) and a verb (**blew**).

The tent blew down **because it had been windy**.

main clause — The tent blew down
conjunction — because
subordinate clause — it had been windy. Tells us *why* the tent blew down. Does not make sense by itself.

Clauses are often joined by a conjunction.

Although his father was a footballer, Oliver did not like sport.

conjunction — Although
subordinate clause — his father was a footballer. Can come first. Does not make sense by itself.
main clause — Oliver did not like sport. Does not have to come first. Makes sense by itself.

✔ Skills Check

1. Underline the main clause in each sentence.

a. Although they read all the information, the committee decided to close all the libraries.

b. There was a flood warning because it had rained a lot.

c. After eating the meal, James was full!

2. Underline the subordinate clause in each sentence.

a. The dragon roared as he blew flames from his mouth.

b. Despite reducing the price, the house didn't sell.

c. School had a cake sale that did very well.

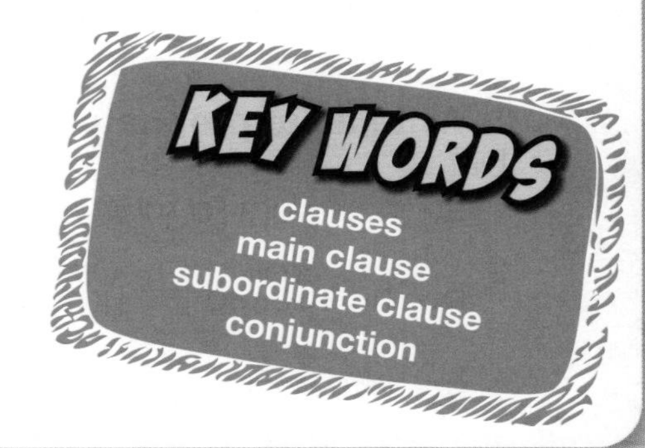

Co-ordinating conjunctions

What is a co-ordinating conjunction?

Recap

A **co-ordinating conjunction** joins two clauses which would make sense on their own.

Revise

An easy way to remember the co-ordinating conjunctions: the initial letters spell fanboys!

The co-ordinating conjunctions are:

for | and | nor | but | or | yet | so

co-ordinating conjunction

The football team trained hard **but** they weren't winning any matches.

Each part of the sentence makes sense by itself.

co-ordinating conjunction

Josh could go to the cinema **or** he might go bowling.

Each part makes sense.

KEY WORDS

co-ordinating conjunctions

✔ Skills Check

1. Choose the best conjunction for each sentence.

and | or | but

a. We could have a barbecue ______________ we could eat inside.

b. The carpet is dirty ______________ we can clean it.

c. I am playing netball ______________ I want to be on the team.

2. Write an appropriate final clause for these sentences.

a. They went to town so ____.

b. She stroked the goat yet ____.

c. I love watching the swans for ____.

Subordinating conjunctions

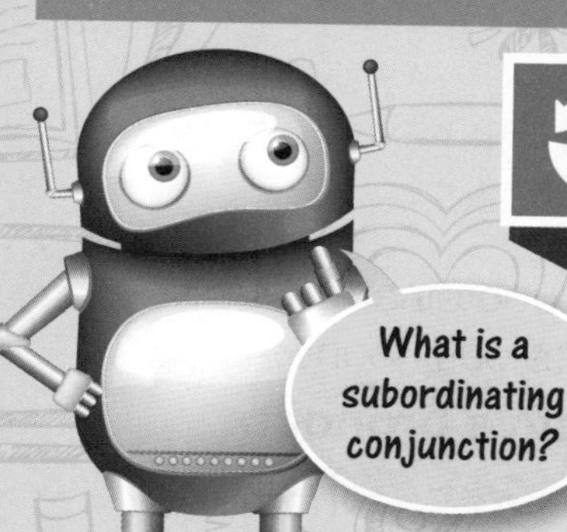
What is a subordinating conjunction?

Recap

A **subordinating conjunction** introduces a subordinate clause, which is dependent on the main clause. Subordinating conjunctions include:

because | if | when | since | before | that | then

Revise

subordinating conjunction

I can't sleep **because** it's so noisy!

main clause — subordinate clause

subordinating conjunction

I haven't seen you **since** I saw you last week!

main clause — subordinate clause

A subordinate conjunction introduces a subordinate clause.

KEY WORDS
subordinating conjunctions

✓ Skills Check

1. Use each conjunction once to join these sentences.

because | if | then

a. I wanted to go camping ______________ it was sunny.

b. I ran downstairs ______________ the doorbell rang.

c. The door creaked open, ______________ a hand appeared.

2. Put a tick in each row, to show whether each sentence uses a co-ordinating or subordinating conjunction.

Sentence	Co-ordinating conjunction	Subordinating conjunction
I met a school friend **when** I was leaving the library.		
We went bowling **and** we went for a pizza.		
You can eat some cake **if** you are hungry.		
Would you like to play this game **or** play your new game?		

Relative clauses

Recap

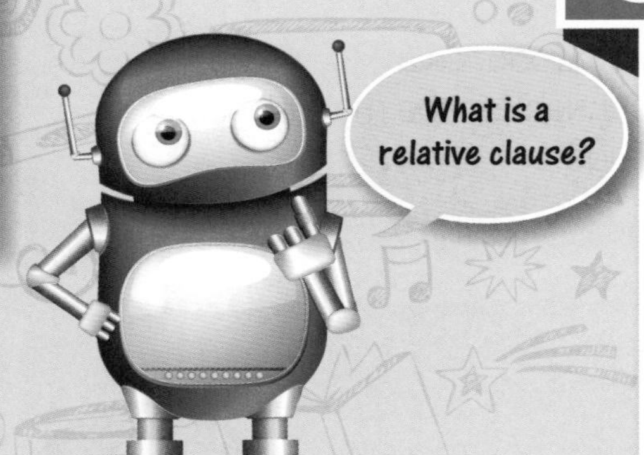

A **relative clause** is a type of subordinate clause that adds information about a previous noun.

Relative clauses start with a **relative pronoun**:

that **which** **who** **whom**
whose **where** **when**

Relative pronouns introduce a relative clause and are used to start a description about a noun.

Revise

Relative clauses describe or modify a noun.

The **dog**, **which was barking**, wanted to go out.

↑ **Relative clause**, starts with **which**. Describes what the **dog** was doing. It modifies the noun.

The **woman**, **who was very old**, walked with a stick.

↑ **Relative clause**, starts with **who**. Describes the **woman**. It describes the noun.

The relative pronouns:

- **who**, **whom**, **whose** refer to people
- **which**, **that** refer to things
- **when** refers to time
- **where** refers to places.

Relative clauses are usually enclosed by commas. They start with a relative pronoun.

✔ Skills Check

1. Underline the relative clause in each sentence.

a. The weather forecast, that we were listening to, told us there would be snow.

b. The man, whose window it was, said it would need to be repaired.

c. The pitch, where the game was to be played, was waterlogged.

2. What does the pronoun 'which' refer to in this sentence?

The sofa, which needed re-covering, was very comfortable.

Personal and possessive pronouns

Recap

A **pronoun** replaces a noun. There are different types of pronoun. **Personal** and **possessive pronouns** are used to replace people or things.

Revise

The personal pronouns are:

I | you | she | he | it | we | they

Ellie went to the park and **she** went on the swings.

Ellie is replaced by the **pronoun she** in the second clause.

They are reading a book.

The **pronoun they** refers to a group of people.

There are also the possessive pronouns:

mine | yours | hers | his | its | ours | theirs

Jack gave me **his** stickers.

The **pronoun his** replaces **Jack**, to avoid repetition.

KEY WORDS
pronouns
personal pronouns
possessive pronouns

Tips

male name → male pronoun
he | his

female name → female pronoun
she | her

neutral (not male or female)
it | its

plural names /objects → plural pronoun
they | their

✓ Skills Check

1. What does the pronoun 'it' refer to in this sentence?
The hotel borders a beautiful sandy beach and it offers great luxury.

____.

2. Choose the best pronoun for each sentence.

his | their | my

a. Oscar played with ______________ toy engine.

b. I couldn't wait to open ______________ presents.

c. The children enjoyed ______________ swim.

Prepositions

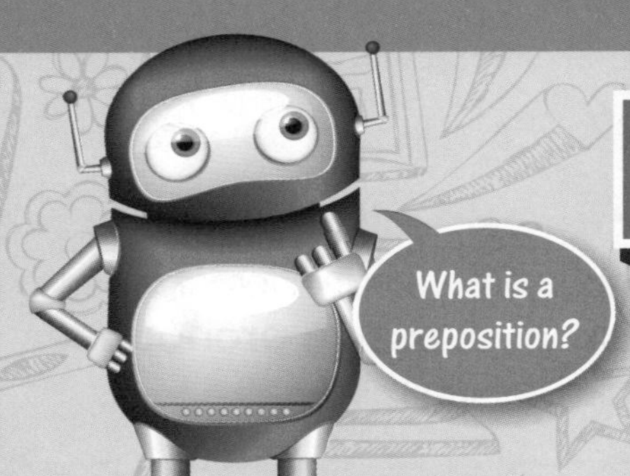

What is a preposition?

Recap

A **preposition** links nouns, pronouns or a noun phrase to another word in the sentence.

KEY WORDS
prepositions

Revise

Here are some common prepositions:

about	above	across	after	around	as	at	before
behind	below	beneath	beside	between	by	for	from
in	in front of	inside	into	of	off	on	onto
out of	outside	over	past	under	up	upon	with

Prepositions often tell us the position of a person or object.

The **dog** was **beside** its basket.

Preposition beside describes the position of the **dog**.

Penny was **inside** the house.

Preposition inside describes the position of **Penny**.

✓ Skills Check

Do not use a preposition at the end of a sentence.

1. Underline the prepositions in these sentences.

a. The marathon runner was under a lot of pressure to finish.

b. We had to queue outside the theatre to get tickets.

c. Aisha was between Orla and Gita.

2. Write a sentence using the preposition 'beneath'.

___

Determiners

What is a determiner?

Recap

A **determiner** is used to define an object or person (a noun).

Revise

Let's look at the different types of determiner.

Articles	Quantifiers	Demonstratives	Possessives
the, a, an	All numbers: one, two... Ordinals: first, second... many, some, every, any	this, those, these	my, your, our, his her, their

These are just some examples – there are others.

There were **many** ducks on **the** water and **my** gran gave me **some** bread to feed them.

↑ **quantifier** ↑ **article** ↑ **possessive** ↑ **quantifier**

Each determiner defines the noun that follows it:

my gran (not anyone else's)

many ducks (not one or a few)

You don't need to know the names of each type of determiner, though it will help to be aware of them.

✓ Skills Check

1. Underline all the determiners in each sentence.

a. An icy wind blew and many people were hurrying back to their homes.

b. Our accommodation was a disappointment and we telephoned its owner.

c. Jane arranged lots of tables around the garden and waited for her guests to arrive.

2. Choose the best determiner for each sentence.

first that our

a. I wanted to buy ______________ pair of shoes.

b. We need to pack ______________ cases.

c. It was her ______________ time at gymnastics.

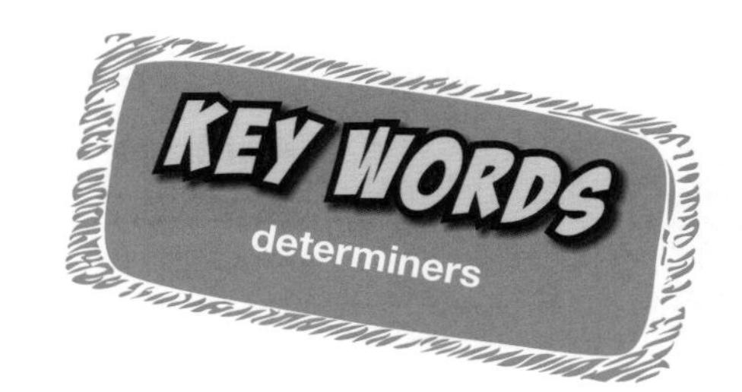
KEY WORDS
determiners

Sentence types: statements and questions

Recap

What are the sentence types?

There are four types of **sentence**: **statements**, **questions**, **exclamations** and **commands**.

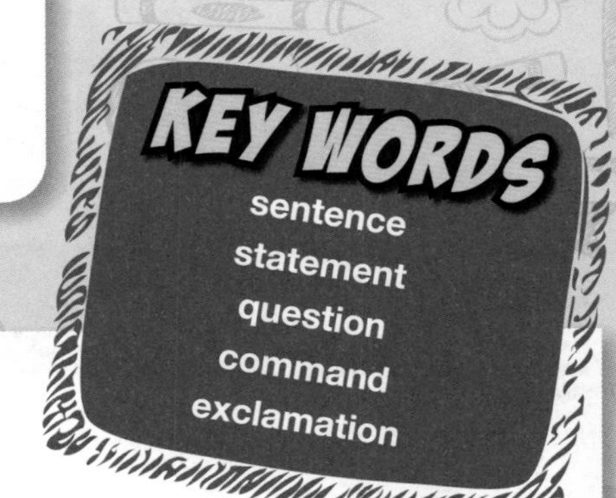

Revise

All sentences start with a capital letter.

A statement: states a fact and ends with a full stop.

London is the capital of England. His name is Josh.

Both state a fact and end with a full stop = **statements**.

A question: asks a question and ends with a question mark.

Who is going to the party? What are you doing?

Both ask a question and end with a question mark = **questions**.

✓ Skills Check

1. Draw lines to match the sentence to the type of sentence.

Where is the nearest petrol station? **Statement**

I wonder where I will find a petrol station. **Question**

2. Write a question starting with the word below.

Who ______________________________

3. Choose the best word to start each question.

When What Which

a. ______________ time do we start school?

b. ______________ is the best way to the beach?

c. ______________ are you going to Scotland?

Questions often start with a question word:

who what
where why
which when

They all start with **wh**!

Sentence types: exclamations and commands

Revise

A command: tells someone to do something and can end with an exclamation mark. It is sometimes called an imperative sentence.

Get off the grass! ↖ Tidy your bedroom! ↗

Both are forceful **commands** and need an exclamation mark

Please get off the grass.

This is not forceful. It is just a polite request. An exclamation mark is not needed.

An exclamation: expresses excitement, emotion or surprise and ends with an exclamation mark.

How fantastic! ↖ What a fierce dog! ↗

Expresses relief or pleasure. Expresses fear or surprise.

Both are **exclamations** and end with an exclamation mark.

✔ Skills Check

Try saying a sentence. Think about what type of sentence it is. Are you asking a question? Do you need to sound forceful or surprised?

1. Put a tick in the correct column to show the sentence type.

Sentence	Statement	Question	Command	Exclamation
Why is the dog barking				
What a beautiful baby				
Line up, quietly				
The chocolate ice-cream was delicious				

2. Insert the correct punctuation in each sentence.

. ? !

a. It was a very exciting game________

b. You had an exciting time at Amelia's, didn't you________

c. Make it more exciting________

d. How exciting________

Apostrophes: contraction

What is an apostrophe for contraction?

Recap

An **apostrophe** for **contraction** is a punctuation mark used to show where letters have been missed out when two words are joined.

KEY WORDS
apostrophes
contraction

Revise

We use the apostrophe to show where letters have been missed out.

he is = he's
missing letter – **i**

he would = he'd
missing letters – **woul**

they are = they're
missing letter – **a**

is not = isn't
missing letter – **o**

The apostrophe must replace the missing letter or letters in the same place.

Tips

Here are some common contractions:

you are	→	you're
did not	→	didn't
was not	→	wasn't
could not	→	couldn't
I will	→	I'll
we will	→	we'll
cannot	→	can't
I have	→	I've

Exception to the rule: **will not → won't**

We often join two words together when speaking or writing informally. Try to work out what the original two words were. Which letters have been missed out? Where should the apostrophe go?

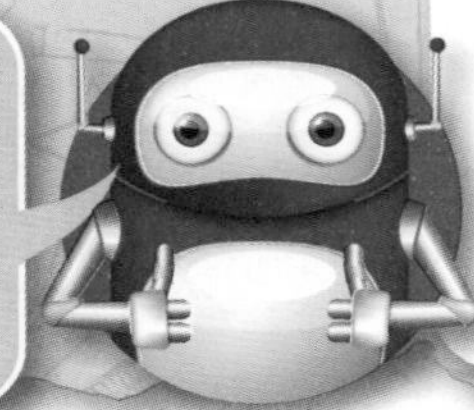

✔ Skills Check

1. Underline the contraction in each pair which has the apostrophe in the correct place.

a. w'ed   we'd

b. wouldn't   would'nt

c. theres'   there's

2. Write each contraction in full on the line below.

a. The **weather's** cold tonight.

____

b. **It's** a long time before **they'll** arrive.

_______________ _______________

c. I **should've** finished it.

____

Apostrophes: possession

Recap

What is an apostrophe for possession?

An apostrophe and the letter **s** are often used to show **possession**; to show when an object belongs to someone or something.

Revise

KEY WORDS
possession
plural
singular

To use an apostrophe to show possession you need to know if the possessor of the object is **singular** or **plural**. This will help you decide where to put the apostrophe.

Single possessor

the girl's bike
↑
one girl: apostrophe + s

the dog's waggy tail
↑
one dog: apostrophe + s

Plural possessors

the girls' bikes
↑
several girls: s + apostrophe

the dogs' waggy tails
↑
several dogs: s + apostrophe

Check how many possessors there are.
One possessor = apostrophe + s
Several possessors = s + apostrophe
Watch out for apostrophes with irregular plurals: children's sheep's

Skills Check

1. Insert apostrophes in the correct places to show possession.

a. Pippis food bowl was empty.

b. The childrens outing was very successful.

c. The swans care of their cygnets was very touching.

2. Change the underlined words to plurals and insert apostrophes in the correct place. Write the new sentence.

The fairy's dress shimmered in the candle's glow.

____

____

Tips

Before adding an apostrophe, be sure that you need to show possession.

The girls went on a long journey.
↑
Several girls – no possession.

The journey belongs to the girls – possession.
↓
The girls' long journey gave them a chance to chat.

Commas to clarify meaning

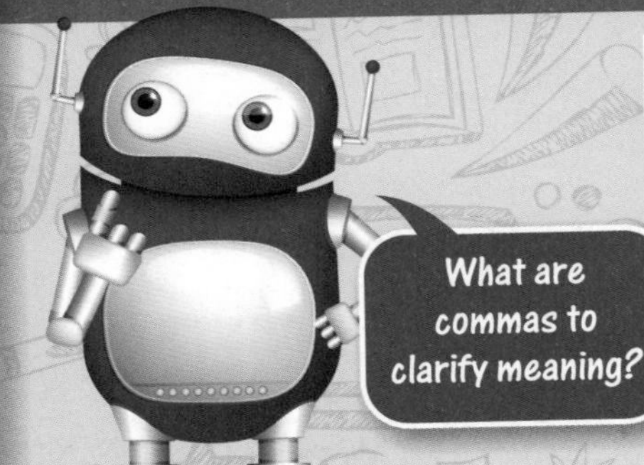
What are commas to clarify meaning?

Recap

Commas are placed in sentences to help us understand the meaning. Using commas within a sentence, can help make the meaning clearer and avoid ambiguity.

Revise

Sometimes the meaning isn't clear without commas.
In the following sentences the words are the same but the comma makes the meaning different:

"Can we go to see Gran?"	Someone is asking if we can go to see Gran.
"Can we go to see, Gran?"	Gran is being asked if we can go to see something.

The comma alters the meaning.
In the next two sentences, the comma alters the meaning again in the same way.

"My mother says Shona is beautiful."	My mother is saying that Shona is beautiful.
"My mother, says Shona, is beautiful."	Shona is saying that my mother is beautiful.

✓ Skills Check

KEY WORDS
commas

1. Explain the meaning of these sentences.

a. "Tell your cousin Alex."

____

b. "Shall we eat Donna?"

____

2. Place commas in each sentence to make the meaning clear.

a. "Tell your cousin Alex."

____

b. "Shall we eat Donna?"

____

Commas after fronted adverbials

What are commas after fronted adverbials?

Recap

A **fronted adverbial** is an adverb or an adverbial phrase, which is at the beginning of a sentence. A fronted adverbial is always followed by a comma.

Revise

After the monsoon, the sun dried up the ground.

↗ fronted adverbial ↖ comma

During winter, roads are often blocked by snow.

↗ fronted adverbial ↖ comma

✔ Skills Check

KEY WORDS
fronted adverbials

1. Place commas in the correct places in these sentences.

a. At the end of the street there is a sweet shop.

b. Tomorrow night there will be a full moon.

c. Poorly cooked the food was inedible.

2. Write your own fronted adverbials, with the correct punctuation, at the start of these sentences.

a. ______________________________ you'll find the treasure.

b. ______________________________ we will go on holiday.

c. ______________________________ the tent blew away.

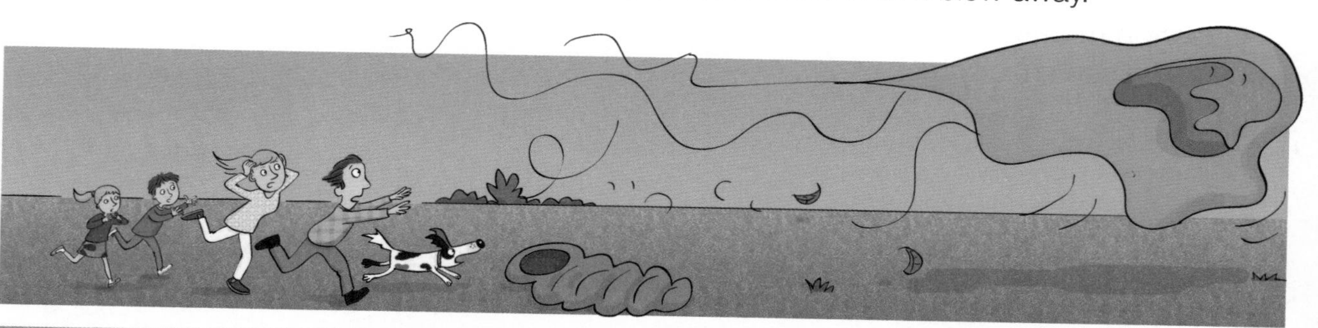

Inverted commas

Recap

Inverted commas are also called 'speech marks'. They go around **direct speech** to show what is being said.

Revise

Inverted commas go at the beginning and end of speech.

"How much is that dress?"

inverted commas → what is being said → **inverted commas**

Inverted commas *always* include one of the following punctuation marks:

comma | **full stop** | **question mark** | **exclamation mark**

The punctuation marks always come between the last word and the second set of inverted commas.

"There's a monster in the cupboard!"

inverted commas → what is being said → **punctuation** → inverted commas

A **comma** is used when the writing continues past the end of the speech.

"This is the boy**,**" said the teacher.

A **full stop** is only used when the speech is the end of the writing. In this case, the comma moves in front of the first set of inverted commas.

The teacher said**,** "This is the new boy**.**"

Question marks and **exclamation marks** are used in the same way depending upon the sentence types.

The teacher said, "Is this the new boy**?**"

The teacher said, "This is the new boy**!**"

✔ Skills Check

1. Place inverted commas in the correct places in the following sentences.

a. Today is Monday.

b. How are you?

c. Stop!

d. Have you finished your work? asked the teacher.

e. The teacher asked, Have you finished your work?

Monday	Saturday
Tuesday	
Wednesday	Sunday
Thursday	
Friday	

2. Rewrite the following sentences with inverted commas and the correct punctuation.

a. The teacher looked at the boy and said Well done!

____

b. It will rain tomorrow said the weather forecaster.

____

c. Look out shouted the driver

____

d. We have some orange juice we also have some mango juice said the waiter

____

____

- Everything that is being said **and** a punctuation mark goes inside the inverted commas.
- Make sure you use the correct punctuation mark **before** the second set of inverted commas.

Remember the comma after words like said when you are using inverted commas.

Parenthesis

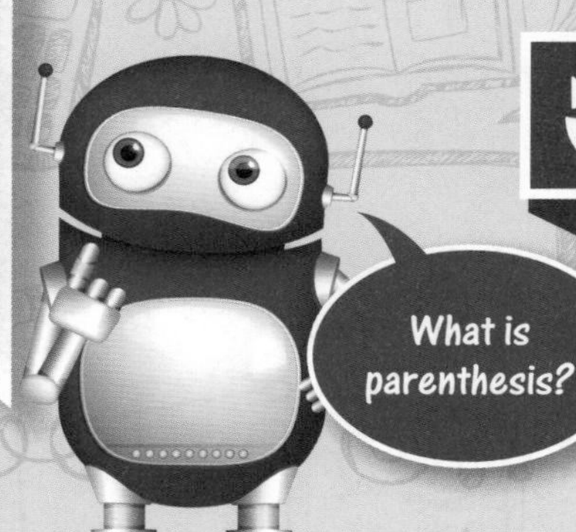

What is parenthesis?

Recap

Parenthesis is the term used for a word, clause or phrase that is inserted into a sentence to provide more detail.

- Parenthesis is what is written inside **brackets**.
- **Commas** and **dashes** can do the same job as brackets.

Revise

The following sentence gives a small piece of information:

My sister is getting married next week.

By adding parenthesis, more detail is given but the meaning remains the same:

My sister (who is older than me) is getting married next week.

parenthesis with brackets

Commas and pairs of dashes can do the same job as brackets:

My sister, who is older than me, is getting married next week.

parenthesis with commas

My sister – who is older than me – is getting married next week.

parenthesis with dashes

Dashes tend to be used in less formal writing, such as in an email.

Remember, parenthesis is the information you add, not the punctuation around it.

✔ Skills Check

1. a. Insert the parenthesis into the following sentence, using brackets.

Toby was lost for seven days. **Parenthesis**: *a six-year-old collie dog*

____

b. Insert the parenthesis into the correct place in the following sentence, using commas.

There are many ways to climb Mount Snowdon. **Parenthesis**: *most of them difficult*

____

c. Insert the parenthesis into the correct place in the following sentence, using dashes.

'Grab a piece of my heart' will be number one next week. **Parenthesis**: *such a great song*

____

d. Insert Parenthesis 1 and Parenthesis 2 into the correct places in the following sentence using dashes and commas.

My new book is going to be a best seller.

Parenthesis 1: *Wheelchair Warrior*

Parenthesis 2: *according to my publisher*

____

____

2. Rewrite this sentence without the parenthesis.

Our favourite place – it's so romantic – is Venice.

____

Paragraphs

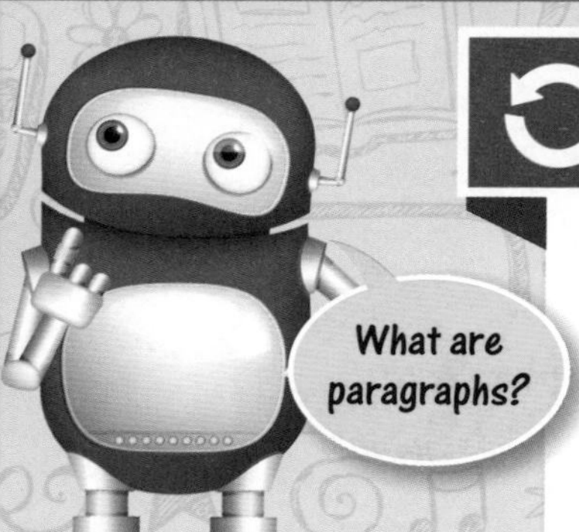

Recap

Paragraphs organise writing to make it easier to understand.

- They break text down into small sections so it is easy to read.
- They are a series of sentences about the same idea.
- We start a new paragraph for each different idea, place, time, character or event.

Revise

In the following story, Jumila has been to see one of her friends.

At the end of the street, Jumila hesitated. She could see her house in the distance. She forced herself to move forwards.

new paragraph because Jumila has moved to a different place

When she reached the gate, Jumila stopped again. This was not going to be easy. She waited nervously for a few seconds before pushing the gate open.

new paragraph | different place

Jumila's father waited inside the house. He looked at his watch. What was keeping Jumila?

different character | different idea

Skills Check

1. **a.** Rewrite the following as two paragraphs.

   Jumila walked slowly towards the door of the house. She did not know what would happen next. She was late and she knew it. Ten seconds later she was inside facing her father.

   ____

   ____

   ____

   b. What has changed to require a new paragraph?

   ____

Headings and subheadings

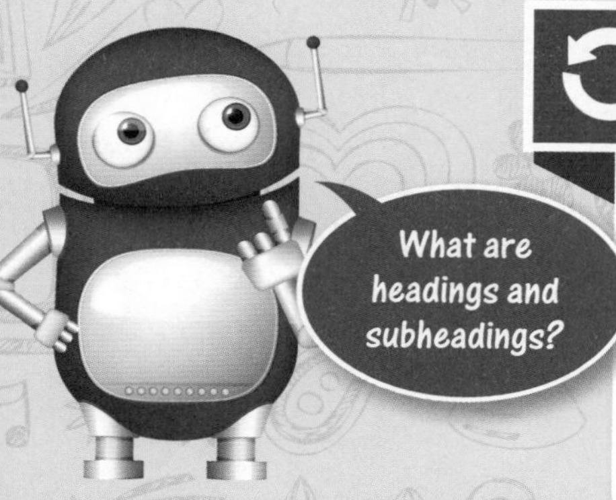
What are headings and subheadings?

Recap

Headings are titles for pieces of writing – they go at the start of the piece.
Subheadings are titles for sections of writing within a longer piece – they go at the start of the section.

- They make the writing easier to read by structuring it.
- They often summarise the writing.

Revise

The North Pole ⟵ **Heading** – tells us what the whole piece is about.

Who lives there? ⟵ **Subheading** – gives a summary of this section.

Despite its ferociously cold temperatures, there is a surprising amount of life around the North Pole. Polar bears and Arctic foxes roam the land whilst in the sea there are whales and seals.

✔ Skills Check

1. Why do we use headings and subheadings?

We use headings ______________________________

We use subheadings ______________________________

2. Read the following article and give a heading and two subheadings.

Heading: ______________________________

Subheading 1: ______________________________

Round about the time your parents were born, nobody could predict the changes in communication technology. Back then, which seems like the Stone Age now, you could only write to someone or telephone them.

Subheading 2: ______________________________

Today, you can still do things the old-fashioned way but you can also call on a mobile phone, video-conference, conference call, text, email or use social media. The future is here now!

Tips

The text in a question will normally be more than one paragraph long. **Read all of it** and decide what the **main idea** is. That will be the **heading**. Then try to give **short summaries** of **each section**. These will be the **subheadings**.

Vocabulary

Prefixes: mis or dis?

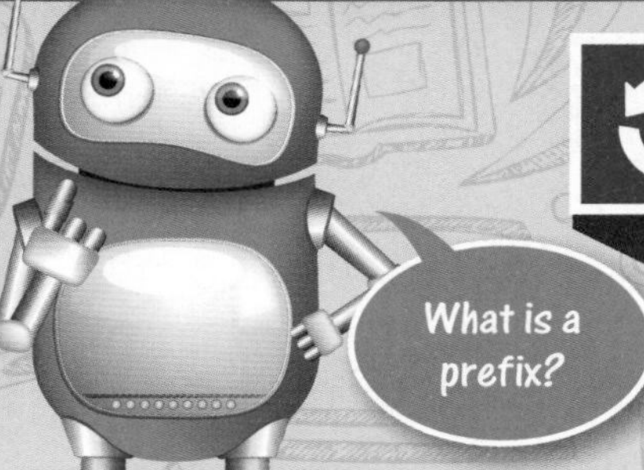

Recap

A **prefix** is added to the beginning of a word to change it into another word, with a different meaning.

Revise

Each prefix has a different meaning. The prefixes **mis** and **dis** both have negative meanings.

dis + **appear** → changes verb to its opposite meaning → **disappear**

dis + **obey** → changes verb to its opposite meaning → **disobey**

mis + **behave** → changes verb to its opposite meaning → **misbehave**

mis + **lead** → changes verb to its opposite meaning → **mislead**

✔ Skills Check

1. Draw lines to join the best prefix to each verb to make a new verb. One has been done for you.

Prefix	Verb	New verb
dis	spell	misspell
mis	appoint	____________
dis	treat	____________
mis	approve	____________

2. Put a circle around the word in each pair which has used the correct prefix.

a. disshapen misshapen

b. disembark misembark

c. dismatch mismatch

d. disbelieve misbelieve

Prefixes: re, de, over

Revise

The prefix **re** means again or back.
It changes the meaning of the verb.

recover	**re**marry	**re**write
to cover again	to marry again	to write again

The prefix **de** changes the verb to its opposite meaning.

decongestion	**de**forest	**de**fuse
to remove congestion	to remove forest	to remove tension

The prefix **over** changes the verb to mean too much.

overact	**over**eat	**over**compensate
to act too much	to eat too much	to compensate too much

✓ Skills Check

1. Choose the best prefix to make a new verb.

re | de | over

a. spend ____________

b. arrange ____________

c. frost ____________

2. Underline the correct word in each sentence.

a. The spy **overcoded** / **decoded** the message.

b. We **declaimed** / **reclaimed** our baggage after the flight.

c. The car **detook** / **overtook** us on the inside lane.

Suffixes: ate

What is a suffix?

Recap

A **suffix** is used at the end of a word, to change it into another word and to change its meaning.

Revise

The suffix **ate** can be added to nouns and adjectives to make verbs.

elastic + **ate** = elastic**ate**
↗ **noun** ↗ **verb**.

The elastic in these trousers has snapped.

I need to elastic**ate** these trousers.

Often, nouns ending **(a)tion** can have related verbs ending with the suffix **ate**.

exagger**ation** → exagger**ate**
accommod**ation** → accommod**ate**
communic**ation** → communic**ate**

Don't forget to drop the final e in a word, before adding an ending!

✔ Skills Check

1. Add 'ate' to change these nouns into verbs. Write the new verb.

a. origin ______________________

b. medic ______________________

c. comment ______________________

2. Use the 'ate' suffix to change these nouns into verbs. Write the new verb.

a. appreciation ______________________

b. domestication ______________________

c. demonstration ______________________

Suffixes: ise, ify

Revise

The suffixes **ify** and **ise** can be added to nouns and adjectives to change them into verbs.
To attach the suffix **ify**:

pure + **ify** = purify
↑
lose final **e**

glory + **ify** - glorify
↑
lose **y**

To attach the suffix **ise**:

apology + **ise** = apolog**ise**
↑
lose **y**

standard + **ise** = standard**ise**
↑
just add suffix

✔ Skills Check

1. Underline the correct form of the word in each sentence.

a. The butter had started to **solidise** / **solidify**.

b. The children were able to **dramatise** / **dramify** the story of Gelert.

c. The farmer needed to **fertilise** / **fertify** his crops.

2. Change these nouns into verbs by adding a suffix. Write the new verb.

a. individual ______________________________

b. quantity ______________________________

c. acid ______________________________

d. terror ______________________________

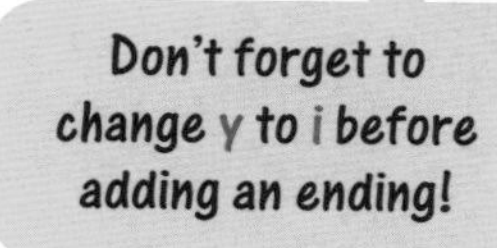

Don't forget to change y to i before adding an ending!

3. Change these adjectives into verbs by adding a suffix. Write the new verb.

a. terrible ______________________________

b. popular ______________________________

c. capital ______________________________

The ise suffix is a lot more common than the ify suffix!

Word families

Recap

A **word family** is a group of words that have a similar feature or pattern. They can often have the same **root word**, but have different beginnings or endings.

Revise

Adding a prefix or suffix will change the meaning of the word and might change its function.
Start with a root word and then try adding different prefixes and suffixes.
How has the word changed?

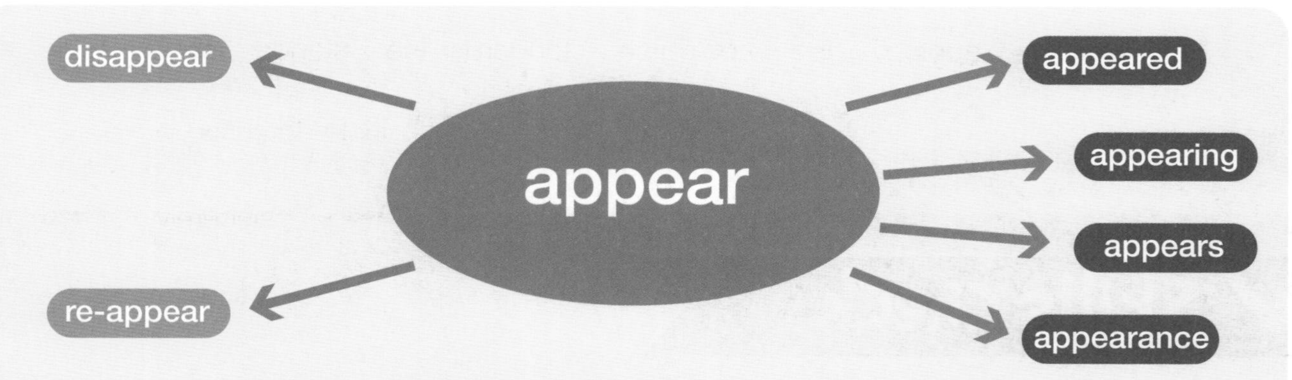

These words all have the root word **appear** but have different beginnings and endings.

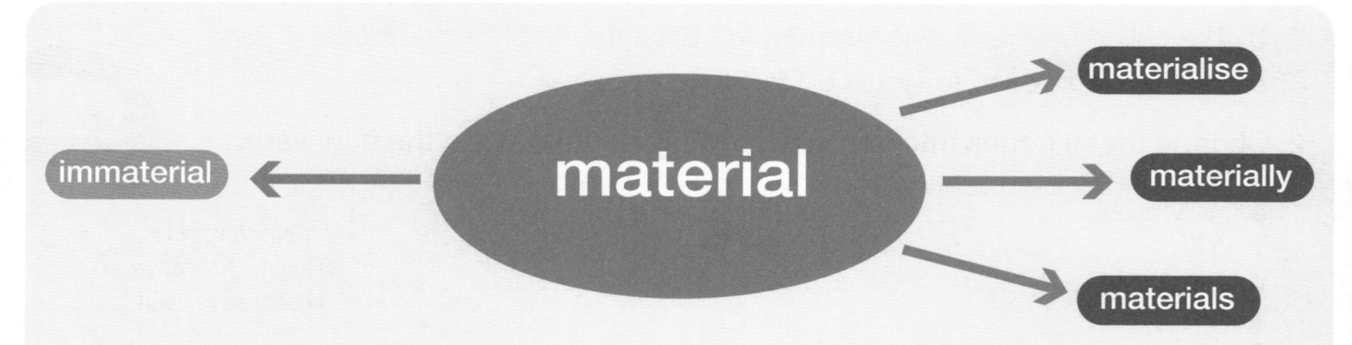

These words all have the root word **material** but have different beginnings and endings.

If you know a root word and can spell it, you can then make lots of other words, using prefixes and suffixes.

✓ Skills Check

1. Can you think of prefixes and suffixes to make new words?

+ prefix	root word	+ suffix
	possess	
	natural	
	remember	
	believe	

2. Colour the root word in each group.

a. accommodation | accommodating | unaccommodating | accommodate

b. insincere | sincerest | sincere | sincerely

c. dissolve | solve | solution | solving

d. continue | discontinued | continuing | continual

3. Underline the word in each group which does not belong.

a. approached | approachable | appearance | unapproachable

b. disinterested | interesting | uninteresting | imaginative

c. community | comparing | communicate | communication

d. prejudiced | prejudicial | privilege | unprejudiced

prefixes suffixes

Spelling

Letter strings: ough

Recap

A letter string is a group of letters which make one sound, within a word.
The letters **ough** can be used to make lots of different sounds!

Revise

rough → **uff** sound
plough → **ow** sound (as in cow)
dough → **oe** sound (as in toe)
cough → **off** sound
through → **oo** sound (as in moon)
ought → **or** sound (as in door)

Say the word. What sound does it have? What letters make that sound?

✔ Skills Check

1. a. Say each word and sort the sound it makes into the correct boxes.

although bought fought trough tough
nought enough bough though thought

uff sound	ow (as in cow)	oe (as in toe)	or (as in for)

b. Which word did not go in the boxes? ______________

2. Write an 'ough' word to fit in each sentence.

a. I ______________ I would be able to get there in time.

b. The sea was very ______________.

c. We crawled ______________ the tunnel.

d. The boxers ______________ in the ring.

e. ______________ it was very stormy, we managed to reach port.

ie or ei?

Recap

What sound does ie make?

What sound does ei make?

field → ee sound (as in tree)

eight → ay sound (as in tray)

BUT after a **c**, **ei** makes an **ee** sound as in **ceiling**!

Revise

- In most words, **i** comes before **e**: chief, thief, believe, field.
- After the letter **c**, **e** usually comes before **i**: receive, deceive, conceive, perceive.
- When **ei** does not come after the letter **c**, it usually makes an **ay** sound: vein, weigh, neighbour.

✔ Skills Check

1. Complete these words with 'ie' or 'ei'.

a. w_____ght

b. ____gth

c. ach____ve

d. n____ghbour

e. c____ling

2. Circle the word which does not have the same vowel sound.

a. grief weight shield deceive

b. neigh weigh vein mischievous

c. niece relief priest neighbour

3. Each word has been misspelled. Write the correct spellings.

a. acheeve ____________

b. theif ____________

c. percieve ____________

d. wayght ____________

e. ayght ____________

f. retreive ____________

Does it make an ee or an ay sound?

Tricky words

Recap

A tricky word may have:

- several **syllables**
- an unusual spelling pattern.

Revise

A syllable is a beat in a word. A syllable has at least one vowel.

Let's look at a word with several syllables.

hippopotamus: 5 syllables

- You need to break the word into parts.
- Say each part of the word slowly and clearly.
- Then work out how to spell each syllable.

Ask: How do I make each sound? What choices are there?

Some words have an unusual spelling pattern

y = i (as in m**y**th)

ph = f (**ph**otogra**ph**)

Here is a tricky word with several syllables and an unusual spelling pattern.

temperature: How many syllables?

Ask: **ar** or **er**?

ch sound, but how is it made: **ch**, **sh**, **t**, **j**?

tem**per**a**ture** = 4 syllables

Say it clearly and you can hear the **m**.

Breaking a word into syllables and then working out how to spell each part makes it easier!

✓ Skills Check

1.

Colour each syllable a different colour	What is the tricky bit in this word?
build	
circle	
vehicle	
relevant	
parliament	
environment	
restaurant	

2. Work out what each tricky word is from the definition.

a. A chart showing the days and months in a year. ____________________

b. The group of people who decide how a country is run, headed by the prime minister.

____

c. The ordinal number for 12. ____________________

d. We use it to look up the meanings and spellings of words. ____________________

e. An occupation or trade. ____________________

3. Underline the correct spelling of each word.

a. vejetable	vegetable	vejutable	vegtable
b. regular	regulur	regalur	wregular
c. seperate	separate	ceperate	separait
d. recagnize	reckognise	recognise	reconise
e. familiar	familier	familyiar	familliar

Ask an adult to read some of the tricky words in this section to you. Try to spell them. Look at the words which you got wrong. What was the tricky bit in each case? Try to memorise the words.

Double trouble

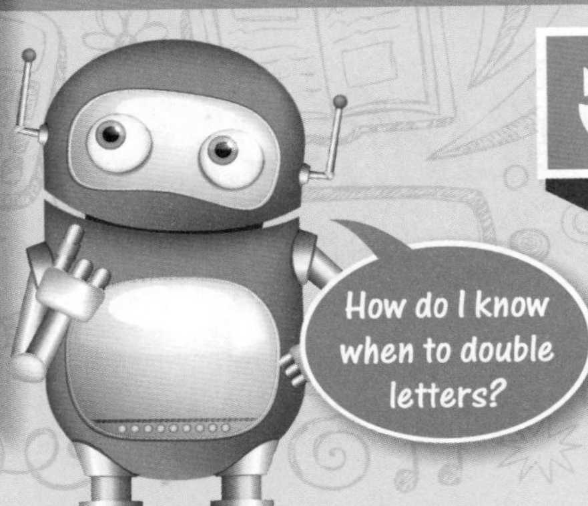

Recap

Some words have double letters in them. It's hard to remember when to double and when not to.

double **d** ↘ double **s** ↙ → address   BUT   double **p** ↘ no double **r** ↙ → appear

Revise

Let's look at words with one pair of double letters.

- double **m** → co**mm**unicate
- double **s** → nece**ss**ary
- double **r** → inte**rr**upt

Learn these words in groups. It helps you to remember them!

Now, we'll look at words with two pairs of double letters.

- double **c**, double **m** → a**cc**o**mm**odate
- double **s**, double **s** → po**ss**e**ss**ion

Some words even have three pairs of double letters.

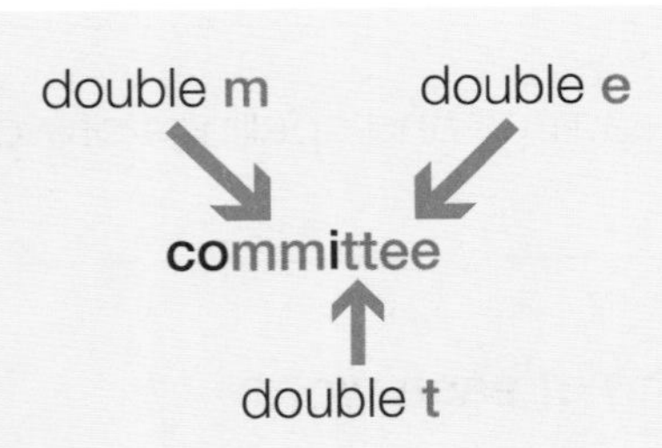

double **m**, double **e**, double **t** → co**mm**i**tt**ee

In words of more than one syllable, a double **consonant** usually shows that the **vowel** before it stands for a short vowel sound. For example: **rattle**, **attached**, **address**.

✔ Skills Check

1. Look at the Word lists for Years 3–4 and Years 5–6 on page 7. Sort the words with double letters into these groups. Now colour the pairs of double letters. One has been done for you.

One pair of double letters	Two pairs of double letters	More than two pairs of double letters
	address	

2. Use words with double letters in place of the words in bold. The words do not need to have the same meaning.

a. **Parallel** lines in a rectangle are equal length. ____________________

b. It was **hard** to jump up that high. ____________________

c. We **often** go abroad for our holidays. ____________________

d. Uma was going to **come with** me on the train journey to London.

____

e. It became **obvious** that they weren't going to finish on time.

____

f. It seemed that the estate agent had **over-stated** the good points in this house! ____________________

Suffixes beginning with a vowel

What is the rule for adding a suffix beginning with a vowel to a word of two syllables?

Recap

- You double the end consonant if the final vowel is stressed.
- You do not double the end consonant if the final vowel is unstressed.

Revise

Below are some examples for adding suffixes to words with two syllables.

for**got** + en = forgotten
stressed vowel sound — suffix begins with a vowel — end consonant doubled

for**get** + ing = forgetting
stressed vowel sound — suffix begins with a vowel — end consonant doubled

be**gin** + er = beginner
stressed vowel sound — suffix begins with a vowel — end consonant doubled

gar**den** + er = gardener
unstressed vowel sound — suffix begins with a vowel — end consonant **not** doubled

Skills Check

Is it a two-syllable word? Is the last vowel sound stressed or unstressed?

1. Add the suffix and write the new word in the third column.

Root word	Suffix	New word
begin	ing	
forbid	en	
regret	ed	
limit	ed	

Adding suffixes to words ending fer

What is the rule for adding suffixes to words ending fer?

Recap

- You double the end consonant if the final vowel is stressed.
- You do not double the end consonant if the final vowel is unstressed.

Revise

Below are some examples for adding suffixes to words ending in **fer**.

refer + ed = referred
stressed vowel sound — end consonant doubled

transfer + ing = transferring
stressed vowel sound — end consonant doubled

BUT

refer + ence = reference
unstressed vowel sound — end consonant **not** doubled

Skills Check

KEY WORDS
suffix
root word
syllable

1. Match each root word to its correct ending.

a. refer + ing — refering / referring

b. transfer + ed — transferred / transfered

c. refer + e — refere / referee

d. prefer + ence — preferrence / preference

e. prefer + ing — preferring / prefering

Suffixes: able and ably

What are the rules for using the suffixes able or ably?

Recap

The suffixes **able** and **ably** are usually used when it is possible to hear the complete root word first.

Revise

The suffixes **able** and **ably** are common.

adore + able = **ador**able
lose final **e** — you can hear the root word

adore + ably = **ador**ably
lose final **e** — you can hear the root word

change + able = **change**able
need final **e** to make soft **g** sound — you can hear the root word

Skills Check

If able is added to words ending ce or ge, keep the final e to make a soft c and soft g sound.

1. Add the suffix 'able' to these words. Write the new word.

a. avail ____________________

b. consider ____________________

c. notice ____________________

d. enjoy ____________________

2. Add the suffix 'ably' to these words. Write the new word.

a. rely ____________________

b. understand ____________________

c. comfort ____________________

d. consider ____________________

Suffixes: ible and ibly

What are the rules for adding ible or ibly to words?

Recap

The suffixes **ible** and **ibly** are usually used when it is not possible to hear the complete root word first.

Revise

The **ible** and **ibly** suffixes are less common.

terror + ible = **terr**ible
lose final syllable — you hear only part of the root word

horror + ibly = **horr**ibly
lose final syllable — you hear only part of the root word

However, there are exceptions!

sense + ible = **sens**ible
lose final **e** — you can hear the root word

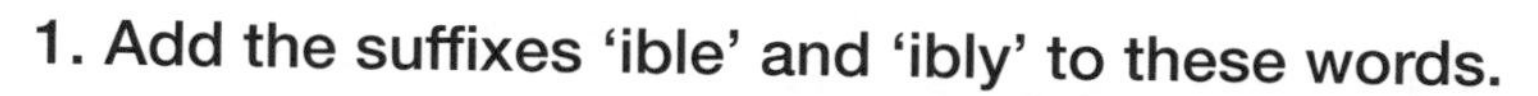

✓ Skills Check

You may only need part of the word.

1. Add the suffixes 'ible' and 'ibly' to these words.

	+ ible	+ ibly
force		
incredulous		
vision		
admission		
comprehension		
response		

Silent letters

When are silent letters used?

Recap

Silent letters are used to write a sound – but you can't hear them when you say the word.

Revise

There are lots of silent letters just waiting to catch you out!
They often pair up with another letter:

wr	has a silent **w**	write wrestle wrought — you only hear the r sound
st	has a silent **t**	listen whistle thistle — you only hear the s sound
kn	has a silent **k**	knowledge knight knee — you only hear the n sound

Skills Check

1. Colour the silent letters in each word.

a. yacht

b. island

c. doubt

d. muscle

2. Underline the correct spelling in each sentence.

a. They rowed the boat towards the deserted **isle** / **ile**.

b. I am going to **rite** / **write** a story.

c. The **lam** / **lamb** was born just after its twin.

d. Dad used the bread **nife** / **knife** to cut me a slice.

e. He cut his **thumb** / **thum** on the glass.

To help you spell a word, pronounce it with the silent letter: **lis – ten**. If you can hear each letter, you will use it when writing the word.

Homophones

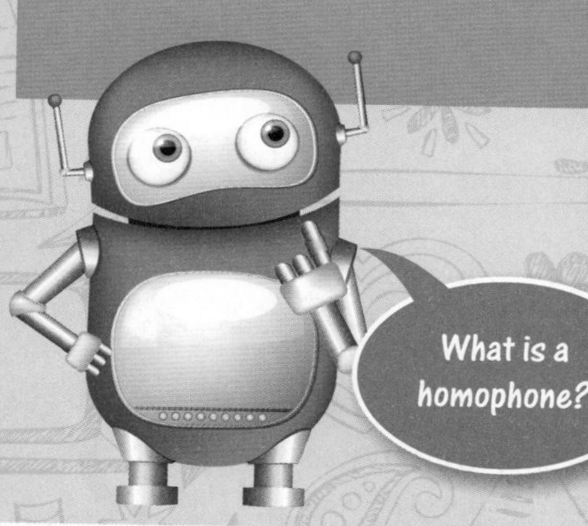

Recap

A **homophone** is a pair of words which sound the same but are spelled differently and mean different things.

Revise

There are lots of homophones.

Look at the **herd** of cows.

I **heard** the birds singing.

Noun or verb?
For words ending **ce** or **se**, the type of word will determine its spelling.

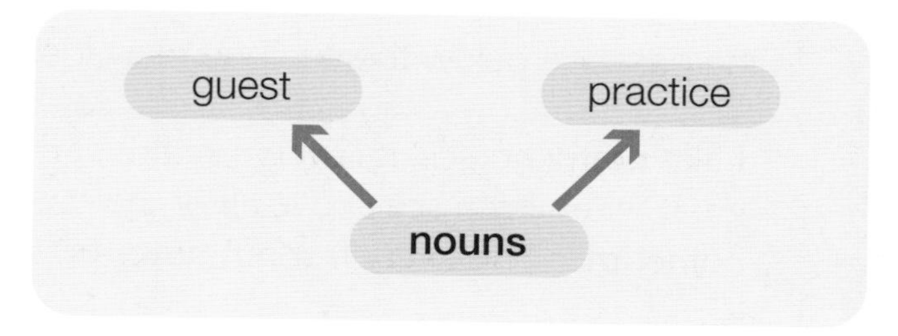

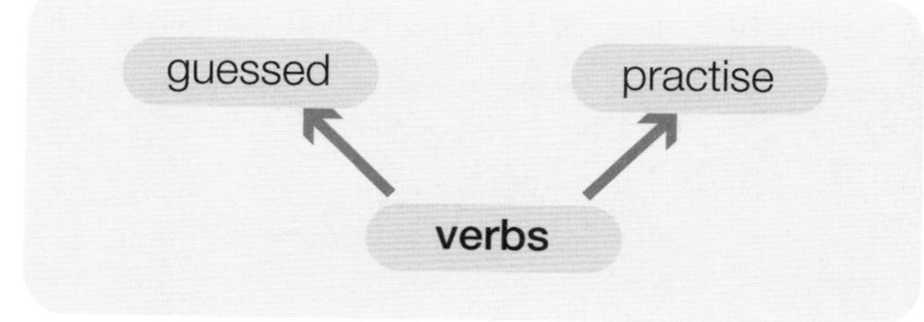

Skills Check

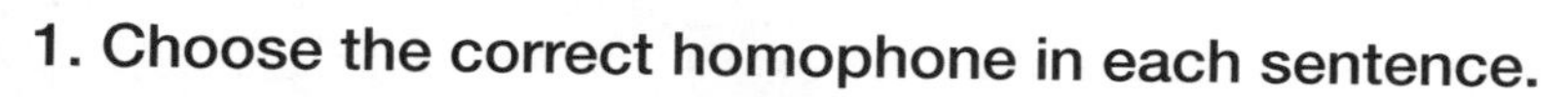

1. Choose the correct homophone in each sentence.

a. I **practised** / **practiced** the piano every day.

b. The bride walked up the **isle** / **aisle**.

c. I prepared the **guest** / **guessed** bedroom for the visitors.

d. He walked straight **passed** / **past** me.

e. **Their** / **They're** / **There** going on holiday next week.

Identifying main ideas

What does identifying main ideas mean?

Recap

The main ideas are the important things the writer is writing about. Often there will only be one main idea in a passage.

When you identify something, you find it in a passage. To find the main ideas, decide what a passage is about overall.

Revise

Don't worry about each individual idea. Look for something that links them all.

In the passage below there is one main idea.

> I like riding horses. It is very exciting. I think it is great when I can see above all of the cars. Sometimes I ride very fast but never on the road.

Each of the sentences is about something different but they are all about horse riding so this is the main idea.

Tips

- Try reading the text and then thinking of a **heading** that fits it overall.
- There are sentences on riding, excitement, seeing over cars and going fast. None of these is the main idea. The link is not horses.
- Each sentence is about **riding**. So the title would be '**Horse riding**'.

Skills Check

Highlight the words in each sentence that show what the sentence is about. Then try to find a link between them.

1. Read this passage and identify the main idea.

> Motor racing is full of danger. From the second the drivers get in their cars, their lives are at risk. Every corner is a hazard. Every bend is a threat to their safety. From the beginning to the end of the race, their lives are in peril.

The main idea is:

____

____

Summarising main ideas

What does summarising main ideas mean?

Recap

Summarise means sum up. When you summarise, you say briefly what the passage is about.
A summary might be one word or a complete sentence. You need to find ideas from the whole text.

Revise

You have to read the whole passage before you can summarise.
In the passage below, there are different ideas for each paragraph.

Life was very different when I was young. We had very little money and couldn't afford to go abroad for our holidays. None of our toys were electronic and computers didn't exist.

← Main idea: what things were like in the past

Nowadays everyone seems much better off. They all go abroad each year. Everyone has a laptop, a tablet and a games console.

← Main idea: what things are like now

There are sentences about the past and the present. The link between the two ideas is the difference. Put this together to summarise the main ideas of the paragraphs: the difference between the past and present.

✓ Skills Check

1. Read the passage below. Fill in the main ideas for each paragraph.

Our cat has really annoyed my dad. He caught her scratching the wallpaper in the hall. He has only just decorated so he isn't very pleased.

← Main idea ____________________
____

My dad isn't very pleased with my mother either. She laughed when she saw what the cat had done!

← Main idea ____________________
____

2. Sometimes, the summary is in the form of a heading or subheading. What do you think a good heading for the passage would be? Tick one.

Our cat ☐ Mum ☐

Dad's decorating ☐ Dad's annoyed! ☐

Identifying key details

What does identifying key details mean?

Recap

- Identify means find.
- The main ideas are the important things that the author wants the reader to know.
- The key details are what the author writes about the main ideas.

Revise

Start by identifying the main idea or ideas.

> There is nothing like a day at the seaside. You can walk along the sea front; play on the beach; risk your fortune in the arcades; ride donkeys or just soak up the sun.

← Main idea: there is nothing like a day at the seaside.

Next highlight the points that tell us more about the main idea.

> **There is nothing like a day at the seaside.** You can walk along the **sea front**; play on the **beach**; risk your fortune in the **arcades**; ride **donkeys** or just soak up **the sun**.

← Each point tells us something different.

Now, use your highlighted points to give three reasons why there is nothing like a day at the seaside.

✓ Skills Check

Remember to highlight the points that tell you more about the main idea

1. Read the passage below.

> Weekends are wonderful. No school. No work. Nothing to do. Just 48 hours of selfish laziness. I usually get up late. Then I meet my mates and go into town. There's no rush and we can do what we want.

a. What is the main point?

____

b. Give three reasons from the text to support this.

1. ____

2. ____

3. ____

Predicting what might happen

What does predicting mean?

Recap

When you predict you say what you think is likely to happen. Usually you have to give reasons for your ideas. These come from clues that are written in the text. This will usually come from fiction texts.

Oh, I see. It means read the story and say what you think will happen next! This is like being a detective.

Revise

Read the following passage and look for the clues to what might happen at the end of it.

> I was driving too fast. It was my own fault. I should have taken more care. When I approached the bend I started to brake. Nothing happened! The road went to the left. I went to the right. All I could see was a duck pond in front of me.

What do you think would happen next? Your answers have to be likely and realistic. A Martian spaceship could arrive and beam the driver up, but is it likely? It's probably not the right answer!

Highlight the important clues.

> I was **driving too fast**. It was my own fault. I should have taken more care. When I approached the bend I started to brake. **Nothing happened! The road went to the left. I went to the right.** All I could see was a **duck pond in front of me.**

It's likely that the car will end up in the pond. Why? The car can't stop and the pond is in front of it. What happened to the driver is anyone's guess.

✔ Skills Check

1. Read this passage. Highlight the important clues.

> I sat in the pond as the water rose around the car. Luckily, a local farmer had seen the accident and had driven across the field in his tractor. I asked him if he could help. He nodded, went back to his tractor and returned with a large piece of rope, which he waved in my direction.

a. What is likely to happen next?

____

____

b. Explain why you think this is likely.

____

Only highlight the points that give clues about what might happen. Use them to make your prediction.

Themes and conventions

What are themes and conventions?

Recap

- Themes are ideas that go throughout the text.
- Conventions are the things that help you know what type of writing it is.

This table shows you some of the themes and conventions.

Type of writing	Possible themes	Convention of this type of writing
Poetry	love, war	verses, rhyme, rhythm, figurative language
Drama	relationships	speech without inverted commas, stage directions
Fiction	myths and legends, adventure, good and evil, loss, fear, danger	heroes and heroines, villains, frightening situations, cliff-hangers, 'good' winning
Non-fiction	history, geography, celebrities, sport, gossip, cars and so on	text books, magazines/newspapers, brochures: headings, subheadings, facts, pictures, columns, bullet points

You need to be able to identify themes and conventions, and comment on them.

That's a lot easier than it sounds! It's a bit like spotting the main ideas but I also need to be able to say how it is written.

Revise

In the passage below, the clues to the **theme** have been highlighted.

> Basitch put a few more logs under her boiling **cauldron**. She watched it bubble and hiss. Two more **toads** and the recipe would be finished. All she had to do then was to utter the **magic words** and the **spell** would be complete!

All of the highlighted words are typical of ones you would find in stories about magic. They are the **conventions**. This is different to the main idea because in this paragraph the main idea would be about making the spell, which is part of the theme of the use of magic.

The theme of the passage below is the triumph of good over evil.
The clues that identify this have been highlighted.

> Basitch hurled her spell at Srodor. He staggered backwards and **collapsed** onto the castle's wall. Basitch cackled loudly **in triumph**. She had **defeated the last of the knights**. Her gloating was suddenly stopped. **Srodor was on his feet!** He grabbed the witch by her shoulders and **heaved her over the wall**, down into the moat where **a moat monster** gratefully accepted its lunch.

To comment on the theme, explain what it is.

For example: It follows the tradition that the heroes always have to seem to be beaten. Despite overwhelming odds, they always win in the end.

To comment on the conventions, show how they help the reader understand the theme.

For example: The text is a fairy story. It has all of the usual elements including a wicked witch, a brave knight, a monster and a final fight.

✔ Skills Check

Highlight the clues to the theme.

1. What is the main theme in this passage?

> Sir Henry knew he would have to be brave. It was no use running away and hiding. He would have to confront the dragon; face its fire; and defeat it, even though he was very scared.

The main theme is ______________________________.

2.

> Pauline set off across the rushing river. The water was soon knee deep and threatening to sweep her away. She forced herself across to the other side and then she set off into the woods. Her torch flickered, fluttered and went out. How would she ever find her way to the treasure now?

a. Find and copy a phrase or sentence that shows that the above passage is an adventure story.

____

b. Explain how your phrase or sentence fits into an adventure story.

____

c. Give two ways that the extract uses the conventions of adventure writing.

____

____

Explaining and justifying inferences

Recap

- Inferences are assumptions that you make from clues in the text. They are the bits the writer doesn't actually tell you.
- Explain means say what you think.
- Justify means give reasons for what you think, using parts of the text to prove your points.

Revise

So this means: Read between the lines; tell us what is happening; and show us why you think that.

Explaining inferences

Some parts of the text below have been highlighted. These are the clues.

> I stopped at the kitchen door and **took my muddy boots off.** I didn't want to get into **trouble again.**

Ask yourself, "What has happened and how do we know?"

- **What do the clues tell us?** The writer has walked into the kitchen before wearing muddy boots and this has caused trouble.
- **What do the clues not tell us?** Who the writer is; why the boots are muddy; what the trouble was.
- **What inferences can we make?** The muddy boots have dirtied the kitchen before. Someone has been annoyed by this.

Justifying inferences

Give reasons for your thoughts. To do this you need proof. This comes from the clues. In the passage above, there are two clues that you can use as evidence:

1. **took my muddy boots off**
2. **trouble again**

To justify the inference, you need to reverse the order of the clues. The most important word is **again**.

This is the key word because it tells us what has happened before and is the basis for the inference (that the writer has dirtied the kitchen before and got into trouble for it).

Now try thinking about what the clues don't tell us:

- who the writer is
- why the boots are muddy
- what the trouble was.

Can you make inferences about those? This is much more difficult as there are no correct answers.

Writing answers

Write down an inference that you can make from the passage.

The writer has been in trouble before for making dirty marks in the kitchen.

Explain an inference that you can make from the passage.

The writer has been in trouble before for making dirty marks in the kitchen. He/she does not want that to happen again so he/she has taken the boots off.

Find and copy two phrases from the text to support your inference.

1. *took my muddy boots off*
2. *trouble again*

✔ Skills Check

1. Read the following passage and answer the questions.

Izzie came home from school and ran straight upstairs to her bedroom. She put her bag on her bed but did not open it. She stared at it, a worried look on her face. Finally, she opened the bag and took her end-of-term report out of it. She hid it under her pillow. She would have to give it to her mother sooner or later but not just yet.

a. How do you think Izzie is feeling?

____

b. Find and copy a phrase that supports your thoughts.

____

c. Do you think the end-of-term report was good or bad?

____

d. Use evidence from the text to support your thoughts.

____

Words in context

What are words in context?

Recap

Words in context means how words are used in the passage.

Revise

When you read the passage, don't just try to work out what the word means. Try to work out what the whole sentence means.

Read the following information from a text about bears.

> Bears have to kill to eat. As a result of this, most people think of bears as being ferocious and frightening.

What does the word 'ferocious' mean in this sentence? Tick one.

- violent ☑
- brave ☐
- strong ☐
- impressive ☐

In this case, you should tick violent as it fits with frightening.

The other answers do not link as well with frightening.

✔ Skills Check

Read the following passage and answer the questions.

> The sea charged in towards the shore. I realised that the sandbank that I was standing on would soon be cut off. I tried in vain to make it back to the safety of the cliff.

1. ***The sea charged in***

Which of these words has a similar meaning to 'charged' in this sentence? Tick one.

- moved ☐
- raced ☐
- poured ☐
- flowed ☐

2. ***I tried in vain to make it back to the safety of the cliff.***
What does 'in vain' mean in this sentence?

____

____

Exploring words in context

Recap

What does exploring words in context mean?

Explore means to go into the meaning of the words.

This is a bit more difficult. Now you have to look at how the words are being used as well.

Revise

To explore, you have to look at a range of possible meanings of a word or phrase. You may need to re-read the sentence or paragraph to work out what the word means.

Read this passage.

> Vietnam is a country of diverse scenery. It has high mountains, stunning beaches and magnificent forests.

Look at this question:

> *Vietnam is a country of diverse scenery.*
> What does 'diverse' mean in this sentence?

If you know what 'diverse' means, that's great. What if you don't? Read the next sentence. Look for a link to help you. **It has high mountains, stunning beaches and magnificent forests.**

The link is 'scenery'. Mountains, beaches and forests are all examples of different scenery. The scenery is 'diverse'. The scenery is *different*.

✓ Skills Check

Read this passage.

> Mauritius is over fifteen hours away by plane. This immense distance does not deter tourists. They go there anyway, despite the hardship.

1. ***This immense distance does not deter tourists.***
Which of these words means the same as 'deter' in this sentence? Tick one.

- help ☐
- discourage ☐
- disillusion ☐
- exhaust ☐

Try all of the words and see which one makes most sense when you read the complete sentence.

2. Explain why you have chosen your answer.

____

____

Enhancing meaning: figurative language

Recap

What is figurative language?

Figurative language is **imagery** used by writers to create word pictures that help the readers *see* what is happening and enhances the meaning.

- Examples of this include: **analogy**, **metaphors**, **similes**, **personification**, **assonance** and **alliteration**.
- You need to write about the effect of the figurative language.

Revise

Figurative language is used a lot in poetry.

Some of the figurative language has been identified in this passage.

> The church was **as quiet as a grave**. Outside **the wind was a wild woman**, whispering **sad sorrows** across the tombstones.

Figurative language	Type of language	Explain the effect
as quiet as a grave	a **simile** – it **compares** by using **'like'** or **'as'**	You often find graves at churches. This simile links the two ideas and shows how quiet it was because graves do not make any noise.
the wind was a wild woman	a **metaphor** – it **compares more strongly** usually using **'is'** or **'was'**	This metaphor helps the reader picture the wind as a mad creature.
sad sorrows	**alliteration** – it creates an effect by repeating consonant sounds, in this case the 's'	It helps the reader hear the noise by repeating the *s* sound which makes a hissing noise like the wind.
wind, wild, woman	**alliteration**	
whispering	**personification** because the wind doesn't actually whisper	

Always say what the effect of the figurative language has been.

KEY WORDS

figurative language
imagery
analogy
metaphor
simile
personification
assonance
alliteration

✓ Skills Check

1. Read the following passage.

> The rain fell like steel spears. I was glad of it because it hid the teardrops that were swimming from my eyes before leaping and diving to the floor in pools of pain. I stumbled home like a blind man.

a. *The rain fell like steel spears.*

The sentence above contains:

Tick **one**.

- a metaphor ☐
- assonance ☐
- onomatopoeia ☐
- a simile ☐

b. Find and copy one example of alliteration from the passage.

____

c. In the table below are examples of figurative language from the passage. In each case, state what type of figurative language is used and explain its effect.

Example	Type/explanation of figurative language
it hid the teardrops that were ***swimming*** *from my eyes*	
in ***pools of pain***	
I stumbled home ***like a blind man***	

How writers use language

How do writers use language?

Recap

Writers use language to have an effect on the reader through:

- vocabulary used
- use of different sentence types and links between them
- different types of text (fairy stories, newspapers, magazines, letters).

You have to write about the effect each has on the reader.

Revise

Always remember to explain why and how the language that is used affects the reader.

Words

Different words show different shades of meaning – like in a thesaurus – only one will be the specific one you want. Imagine, for example, that someone has annoyed you, but how much? Did they **bother**, **upset**, **irritate**, **exasperate** or **infuriate** you? The slightly different meanings help us understand exactly what is meant.

Sentences

Different forms of sentence create a response in the reader.

- **Everybody knows that butter tastes better than margarine.** This is a **statement**. It seems to be a fact but is it? Actually, it's not a fact but it is presented as one so the effect is that the reader believes it is true.
- **Would the world be a better place if everyone tried to be kinder?** This is a **rhetorical question**. It has a clearly expected answer that the reader should agree with.
- **Dev listened carefully to see if the footsteps were still following him. Silence.** The **short sentence** has an effect because of its length. It tells us quickly what we need to know.

Text

Texts are written for different purposes. You need to be able to identify the purpose and show how the writing fits it. This one is written to persuade.

Everyone loves chocolate. Everyone loves milkshakes. Now you can have the best of both with our new Chocolate Milkshake bar. Imagine a chocolate bar that dissolves in your mouth, leaving you with that creamy, frothy, yummy milkshake taste. That's the Chocolate Milkshake bar!

- This includes you! → Everyone
- Can it get better than love?... → loves
- ...it just did. → best of both
- This is what it does. → dissolves in your mouth,
- This is what it tastes like. → milkshake taste.

Read the passage below. It is from a horror story. How do we know? The words that tell you are highlighted.

> A chill wind sliced into Amy's face. The graveyard was no place to be in the dark. Around her feet the soil moved slightly. What could it be? Earth doesn't move by itself. Nothing moves in a graveyard… except ghosts! Ghosts don't exist, do they? Something pushed through the ground. What could it be? What was it? Amy looked in horror. It couldn't be. A finger!

✔ Skills Check

1. The following questions are about the Chocolate Milkshake bar extract.

a. Which word in the sentence 'Everyone loves chocolate' is meant to make the reader think that they should enjoy chocolate too?

____

b. Why have you chosen that word?

____

c. Imagine a chocolate bar that dissolves in your mouth, leaving you with that *creamy, frothy, yummy, milkshake taste*. What is the effect on the reader of the words in italics?

____

____

2. These questions are about the horror story.

Language used	Effect of language
a. A *chill wind sliced into Amy's face. The graveyard was no place to be in the dark.*	Find and copy two phrases that set the scene in these sentences. ______________________ ______________________
b. *Ghosts don't exist, do they?*	What is the effect on the reader of this question? ______________________ ______________________
c. *What could it be? What was it? Amy looked in horror. It couldn't be. A finger!*	Give two ways the writer has tried to build tension in this extract. **1.** ______________________ **2.** ______________________
d. Look at the whole passage.	How can you tell this is from a horror story? ______________________ ______________________

Features of text

What are features of texts?

Recap

- Language features – the way the words are used.
- Structural features – the way the text is organised.
- Presentational features – the way the text looks.

Revise

Here are some examples of different features.

Language features	Structural features	Presentational features
• Figurative language • Short/long sentences • Variety/repetition of words • Rhetorical questions	• Table of contents • Headings and subheadings • Paragraphs or verses	• Pictures and captions • Columns and charts • Text boxes • Fonts and colour

In the following passage, a number of language, structural and presentational features have been identified for you.

heading and bold text → **The New Girl**

All eyes turned like searchlights on the *new* girl.
She felt them burn into her. Alone. She felt so very alone.

simile: eyes turned like searchlights
italics: *new*
metaphor: burn into her
single-word sentence: Alone.

Skills Check

Do you think you're brave?

Would you go out in the dark? Would you go out in a graveyard in the dark? Would you go out in a haunted graveyard in the dark? With no moon? On your own? Alone? Yes? Are you brave? Or stupid?

1. Find and copy examples of language, structural and presentational features in the above text.

Feature	Feature name	Example
Language		
Structural		
Presentational		

Text features contributing to meaning

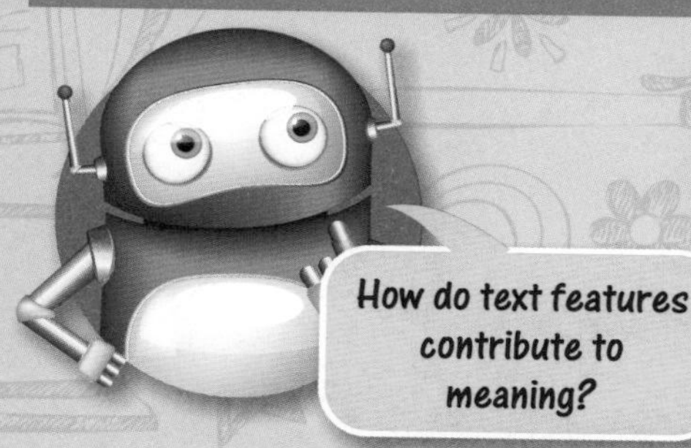
How do text features contribute to meaning?

Recap

Text features are the language, structural and presentational features of texts.
You need to explain how they help the reader understand the meaning of the text.

Revise

This is the second paragraph of **Do you think you're brave?** Some features are highlighted.

subheading and bold text

repetition

Think again
No one near you. No one to hear you. No one to care. Dark. So dark.
Like the dark side of the moon. Are you still feeling brave? Think again.

simile

refers back to subheading

You need to be able to explain how each feature works in the passage as a whole.

Feature	What it does	Explanation
Subheading	Makes it easy to read	Breaks up the text and gives a summary of the main idea of the paragraph
Bold text	Draws attention to important text	Makes it stand out
Repetition	Builds up tension	Emphasises how alone the person is
Simile	Helps the reader imagine the scene	The dark side of the moon has no light; it is in total darkness so nothing can be seen
Ending	Refers back to beginning	Makes the reader answer the question

✔ Skills Check

Read both parts of 'Do you think you're brave?'

1. What does the use of 'you' make the reader feel?

____

2. *Dark. So dark. Like the dark side of the moon.*

How does the writer build up tension in these sentences?

____

Retrieving and recording information

Recap

- Retrieve means find.
- Record means write down.

Look for the first key words in the question: why, what, who, where, when or how.

Revise

Read the following passage. Then look at the example questions.

The Tower of London

Sitting on the north bank of the River Thames, the Tower of London dates back almost a thousand years to William the Conqueror. At times it has been a palace, a stronghold and a prison.

For many years the Tower had a gruesome reputation. At one time, no one wanted to be 'sent to the Tower' because it meant almost certain death. Nowadays it's a tourist attraction. It's still guarded day and night because the Crown Jewels are in it, but it's not frightening any more.

Example questions

What was the reason people did not want to be 'sent to the Tower'?

is the same as

Why did people not want to be 'sent to the Tower'?

Look for the **other key words**: these tell you what to retrieve from the text. In this question they are: 'sent to the Tower'.

Scan the text above for 'sent to the Tower' and you'll find the reason – 'it meant almost certain death'.

Some questions will ask you to tick boxes. For example:

Nowadays, the Tower of London:

Tick **two**.

is on the south bank of the River Thames.	☐
has a gruesome reputation.	☐
is guarded day and night.	☑
is dangerous.	☐
is not frightening any more.	☑

Tips

Look closely to find the answers!
- **Why** = find a reason
- **Who** = find a name
- **Where** = find a place
- **When** = find a time
- **How** = find an explanation
- **What** can be any of the above.

Some questions will ask you to join information together. For example:
Draw lines to link the Tower of London to its uses.

Tower of London

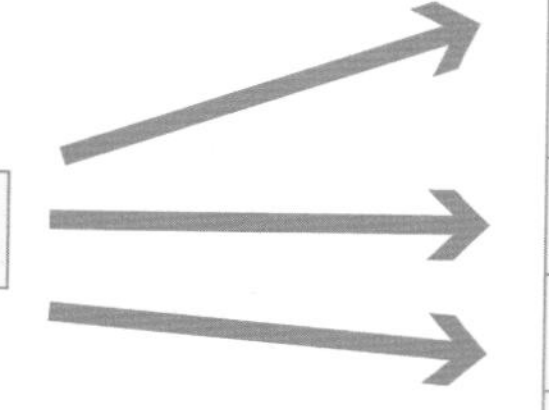

palace
holiday camp
prison
tourist attraction
university

✔ Skills Check

1. Read 'The Tower of London' again.

a. Where is the Tower of London?

____

b. How long ago was the Tower of London first built?

____

c. What kind of reputation did the Tower of London have?

____

d. What is kept in the Tower of London nowadays? Tick **one**.

William the Conqueror ☐
Prisoners ☐
Tourists ☐
The Crown Jewels ☐

Remember to look for the key words. They tell you what to retrieve.

e. Draw lines to link the Tower of London to its history.

Tower of London

has been a stronghold
has been frightening
has always had a good reputation
dates back to Henry VIII
has always been a tourist attraction

Making comparisons

What does making comparisons mean?

Recap

Comparisons show us what is similar or different in a text.

Revise

Read the following passage.

Camping
Love it or hate it, camping is big business.
What's to love?
The outdoor life. The open air. The countryside. The peace and quiet.
What's to hate?
Cold. Wet. Creepy-crawlies. Mud.

Give one reason why people should hate camping and one why people should love it.

Love it: The peace and quiet
Hate it: Mud

Read the continuation of the passage.

If you have to go camping, there are lots of choices you can make.
You could go for a cheap tent. Some even set themselves up. They are waterproof and draft-proof. You can even take a camp bed with you, so they'll be comfortable as well.

Not sure? Well, what about a moderately expensive caravan? They are easy to set up, and in the latest ones your bed's already made. You'll have running hot and cold water and even central heating!

Still not sure? You could always go for a campervan. Smart, elegant and expensive, campervans offer you all the comforts of home while on the move. No setting up. Simply arrive and start your holiday!

To compare, you have to show the similarities and the differences.

Comparison	Tent	Caravan	Campervan
Similarities – cost	Cheap	Moderate	Expensive
Similarities – set-up	Some set themselves up	Easy to set up	No setting up
Differences – comfort	Waterproof and draft-proof	Hot and cold water; heating	All the comforts of home

✓ Skills Check

1. Read the continuation of the camping passage again.

a. Give one thing that is different between caravans and campervans.

____

b. Give one thing that is similar between caravans and campervans.

____

2. Read the following passage.

> Today's music is rubbish: hip hop, garage, house. What does it all mean? Music was better in the 1970s than today: heavy metal, pop, glam rock. Those were the days! The artists then were memorable and their music lives on. Slade, Status Quo, David Bowie and Michael Jackson all continued to have massive careers. Today's nameless, faceless, personality-less, plastic performers will be forgotten instantly. The 1970s had real individual talents. Today's identical stars are mass produced to a formula. Of course, some things never change: record companies will always make the most money.

a. Compare the types of music available today with those of the 1970s.

____

____

____

b. What differences are there between the artists of today and those of the 1970s?

____

____

c. What is similar between now and the 1970s?

____

Fact and opinion

What is fact
and opinion?

Recap

- A **fact** is true and can be proved.
- An **opinion** is what someone thinks or believes.

You need to be able to tell the difference between facts and opinions.

Be careful! Sometimes
opinions are disguised
as facts.

Revise

In the passage below, there is one **fact** and one **opinion**.

> **Henry VIII**
> Without doubt, Henry VIII was the greatest king of England. The changes he made in his lifetime, particularly to the Church, still affect people today.

- **Fact:** The fact is that Henry's changes do still affect us today. **Can this be proved? Yes.** He was the founder of the Church of England. It still exists. Without him, it wouldn't.
- **Opinion:** 'Henry VIII was the greatest king of England.' **Can this be proved? No.** He existed and he was a king, but was he a great king or the greatest? That is a matter of opinion. It's like saying, "Which is the best football team?"

The text makes it seem as if Henry was the greatest king. How? Look at the opening phrase. 'Without doubt' disguises the opinion by making it seem like it is true without giving any evidence. It is possible to measure the tallest, widest, shortest and so on. It is not possible to measure and compare 'greatness'.

To tell if something is a fact, ask the question, 'Can it be proved?'

✔ Skills Check

Read the following passage.

Mount Rushmore National Memorial is one of the most amazing sculptures ever made. Started in 1927 and finished in 1941, the memorial was carved into the granite face of Mount Rushmore in South Dakota. There are four huge heads of presidents of the United States: George Washington, Thomas Jefferson, Theodore Roosevelt and Abraham Lincoln. It was a hard choice! Originally, the figures were meant to be carved from the waist up but the project ran out of money so only the heads were completed. Such a sculpture will never be achieved again.

1. Put a tick in the correct box to show whether each of the following statements is fact or opinion.

	Fact	Opinion
Mount Rushmore National Memorial is one of the most amazing sculptures ever made.		
The memorial was carved into the granite face of Mount Rushmore in South Dakota.		
The project ran out of money.		
Such a sculpture will never be achieved again.		

2. Find and copy three facts from the passage that are not included in the table above.

1. ______________________________

2. ______________________________

3. ______________________________

3. Find and copy one other opinion.

____

Watch out for **'weasel words'**. They're slippery and hard to get hold of. 'Probably' and 'possibly' are two weasel words. They usually tell us that something **is not a fact**.

Maths Made Simple Ages 9–10

Numbers up to 1,000,000

What do the other digits represent?

Recap

23,471 in words is twenty-three thousand, four hundred and seventy-one.

10,000s	1000s	100s	10s	1s
2	3	4	7	1

The **place value** of the **3** digit represents **3000**, the **4** represents **400**.

Revise

1,000,000s	100,000s	10,000s	1000s	100s	10s	1s
	3	4	0	2	6	1

This number is three hundred and forty thousand, two hundred and sixty-one.

Now read these statements aloud. They are both true.

For this symbol: > say *is bigger than* and for this symbol: < say *is smaller than*.

999,999 > 703,374 > 12,029 > 7698

6418 < 30,206 < 163,192 < 1,000,000

Tips

Hey there! Follow my tips and you'll soon find big numbers are not a big problem!

- Write the place-value headings in columns above numbers if you're stuck.
- Use commas in numbers over a hundred thousand (100,000) and numbers over ten thousand (10,000).
- If you get mixed up with the > and < symbols, just think of the symbol as the mouth of a crocodile – the crocodile always eats the bigger number!

876,457  292,345  

Talk maths

Read these numbers and statements aloud.

100,000

1,000,000

1000

10,000

324,492 > 27,920

725 < 4205

500,000

999,999

DID YOU KNOW?

Did you know that 500,000 is half a million?

✔ Check

1. **Write this number in words: 34,805.**

   ____

2. **Write this number in digits.**
   Two hundred and thirty-seven thousand, one hundred and twenty ______________

3. **What does the 4 digit represent in 540,371?** ______________

4. **Put these numbers in order, from smallest to largest.**
   25,612 50,000 725 7 225,421 1,000,000 12 899,372

   ____

5. **Insert the correct sign, < or >, between these pairs of numbers.**

   a. 3521 _____ 5630 **b.** 15,204 _____ 9798 **c.** 833,521 _____ 795,732

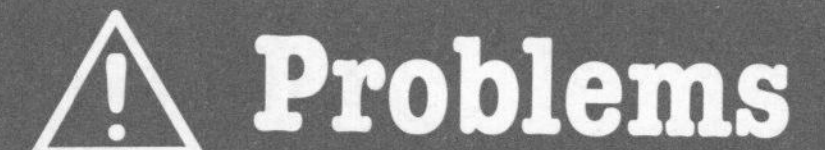

Problems

Football club	Fintan United	Forest Rovers	Winchcomb City
Fans	500,243	96,589	742,104

Brain-teaser Which club has the most fans? ______________

Brain-buster Write the clubs in order, from largest to smallest, according to the number of fans.

______________ ______________ ______________

Counting in steps up to 1,000,000

Can you count on from 0 in 5s? Can you count back from 100 in 10s?

Recap

When we count in steps, we add or subtract the same number each time.

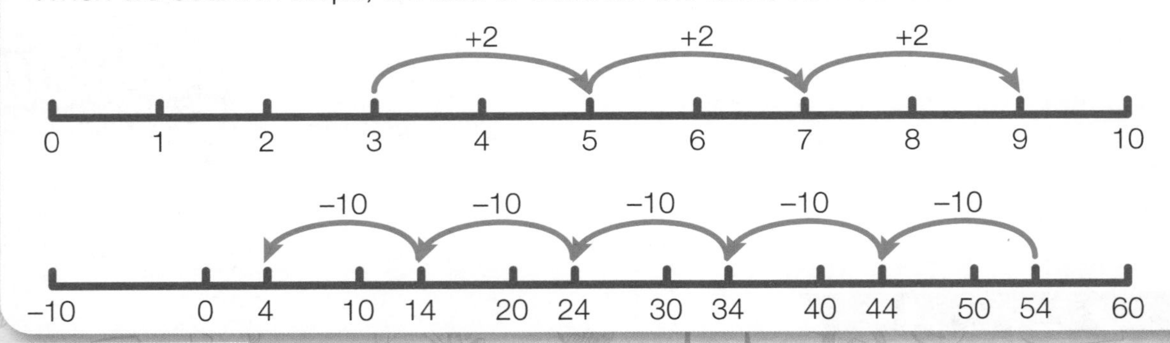

Revise

We can count on or back in steps for any power of 10.

Powers of 10 are numbers that are made by multiplying 10 by 10 a number of times:

- 100 is 10 × 10 or 10 to the power of 2 (102)
- 1000 is 10 × 10 × 10 or 10 to the power of 3 (103)
- 10,000 is 10 × 10 × 10 × 10 or 10 to the power of 4 (104)

This number line goes up to 1000. Starting at 45, we count in steps of 100.

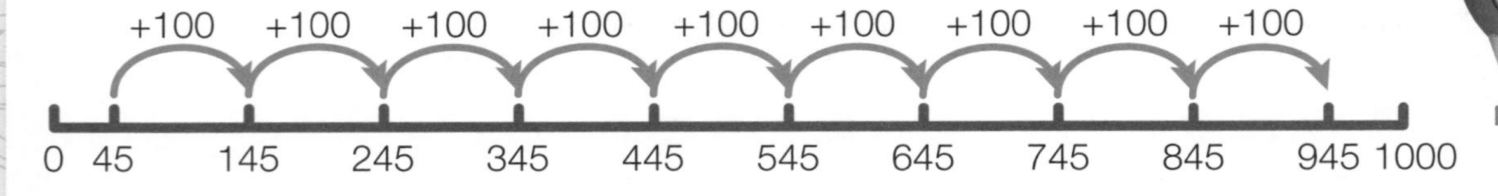

Can you continue the count?

This number line goes up to 60,000. Starting at 1362, we count in steps of 10,000.

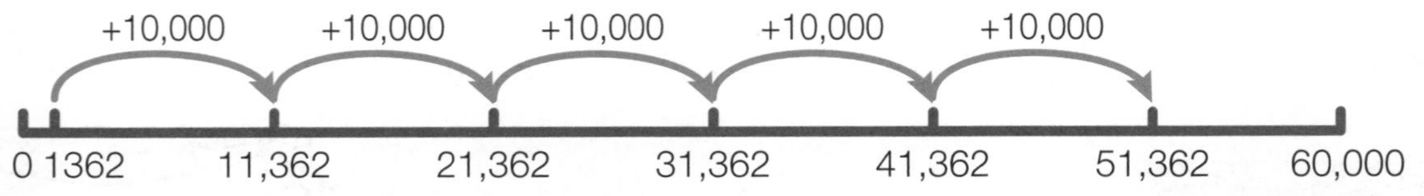

Can you count back in ten thousands? Try starting at 51,362.

- Write the place values in columns above numbers if you're stuck.
- Remember which power of 10 you are adding each time.

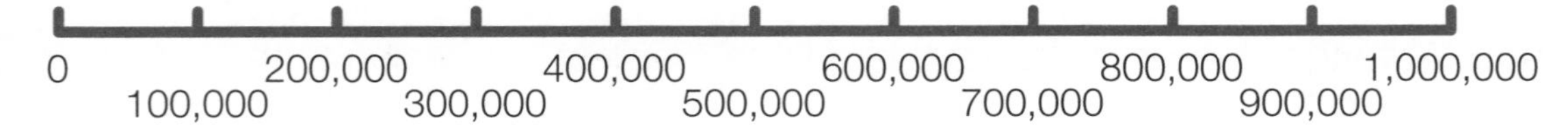

- Choose a number less than 1000, then try counting on in 100,000s.

Count aloud in steps of different powers of 10.

Talk maths

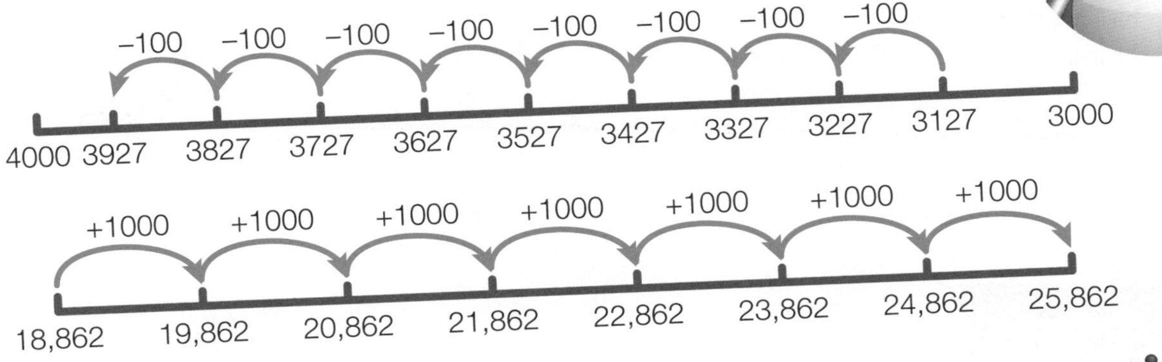

Can you continue counting and say what the next three numbers would be?

✓ Check

1. Complete this sequence, counting on in steps of 100.

   124, ______, ______, ______, ______, ______

2. Complete this sequence, counting back in steps of 1000.

   12,906, ______, ______, ______, ______, ______

3. Complete this sequence, counting on in steps of 100,000.

   320,435, ______, ______, ______, ______, ______

4. Complete this sequence, counting back in steps of 10,000.

   243,000, ______, ______, ______, ______, ______

⚠ Problems

Brain-teaser Evie's mum is saving for a family holiday. She has already saved £746 and will continue to save £100 a month. How many more months of saving will it take before she has over £2000?

______

Brain-buster There are 123,456 people at a football match. At the end of the match 10,000 people leave every ten minutes. How many people will still be in the stadium 50 minutes after the match has finished?

______

Positive and negative numbers

On a number line, when we add numbers we move to the right; when we take away numbers we move to the left.

Recap

Numbers less than zero are called negative numbers.
Numbers more than zero are called positive numbers.

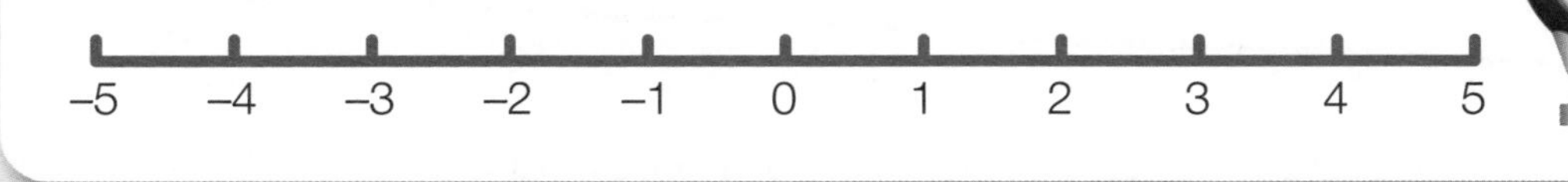

Revise

Temperature is a great way to practise using positive and negative numbers.

- If you start at +5 and count back 5 you stop at 0.
- If you start at –8 and count on 6 you stop at –2.
- If you start at –3 and count on 4 you stop at +1.
- If you start at +2 and count back 10 you stop at –8.

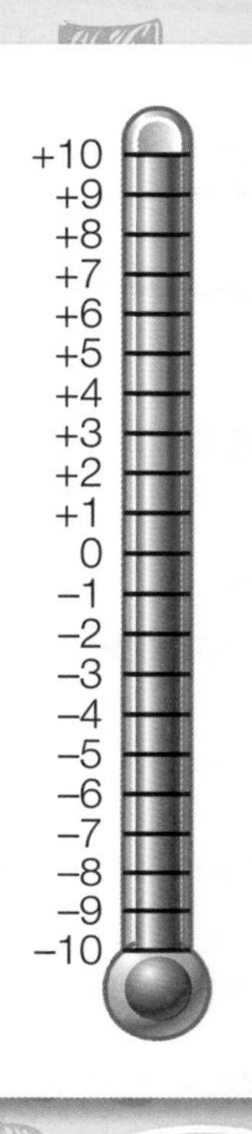

We can do simple calculations with positive and negative numbers.
For example, **2 – 3 = –1** Or, **–3 + 2 = –1**

Try a few of your own and check them with a friend.

Tips

I'm positive my tips will help you!

- Remember that adding, or counting on, moves up a thermometer, or to the right on a number line.

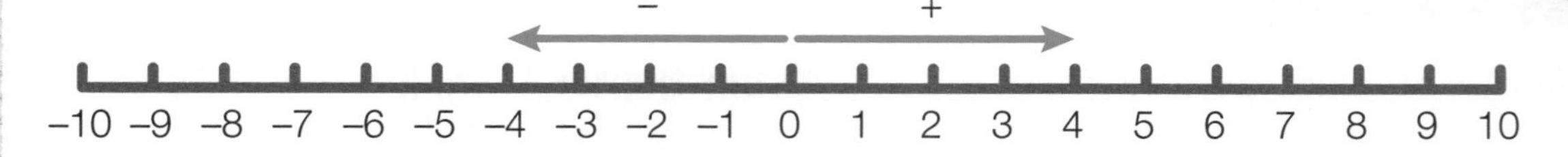

- Use your finger to count on and back from different numbers. Each time say your calculation aloud, such as **minus four add seven equals plus three**.

Talk maths

Make sure you understand the different vocabulary for talking about positive and negative numbers. Read these statements aloud.

We say count on, plus or add.

We say count back, minus or subtract.

We say minus three to subtract three, but we also say minus three to talk about the number −3. We also say negative 3 to talk about the number −3.

We can say positive or plus for numbers greater than zero, but usually we just say the number.

We always say negative or minus for numbers less than zero.

We say plus five to add five, but we also say plus five to talk about the number 5.

✔ Check

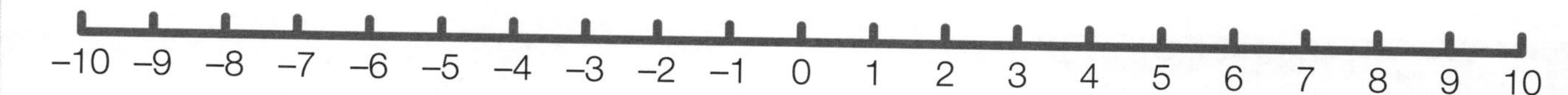

1. Complete these calculations.

a. −5 + 5 = ______ **b.** 5 − 5 = ______ **c.** −2 − 7 = ______ **d.** 3 − 7 = ______

2. Count on from −6 to +6 in steps of 2. Write down each number.

______, ______, ______, ______, ______, ______, ______

3. Insert the missing signs, + or −.

a. 4 ______ 4 = 0 **b.** −5 ______ 6 = 1 **c.** 2 ______ 7 = 9 **d.** 2 ______ 8 = −6

4. Insert the missing numbers.

a. −3 + ______ = 1 **b.** 1 − ______ = −2 **c.** ______ − 7 = 2 **d.** ______ − 10 = −8

⚠ Problems

Brain-teaser The temperature at dusk is 4 degrees Celsius (4°C).
If the temperature drops 6°C by midnight, what will the temperature be? ______________

Brain-buster In January, the temperature at 11am in Montreal, Canada, was −9°C and in Sydney, Australia, it was 27°C. What was the difference in temperature?

____

Number and place value

Rounding numbers

Don't forget, 34 and below round down to 30, 35 and above round up to 40.

Recap

To round a number to the nearest 10 we can look at its position on the number line.

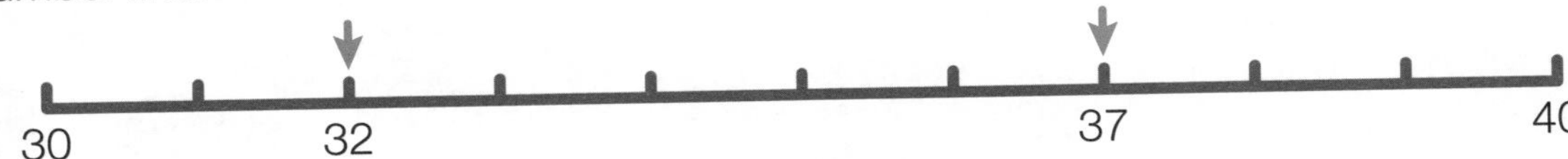

Don't forget (again!), 50 and 500 round up, 49 and 499 round down.

We then look for the nearest 10.

32 rounds down to 30 **37 rounds up to 40**

We can do the same with 100s and 1000s.

355 rounds up to 400 **1268 rounds down to 1000**

355

300 400

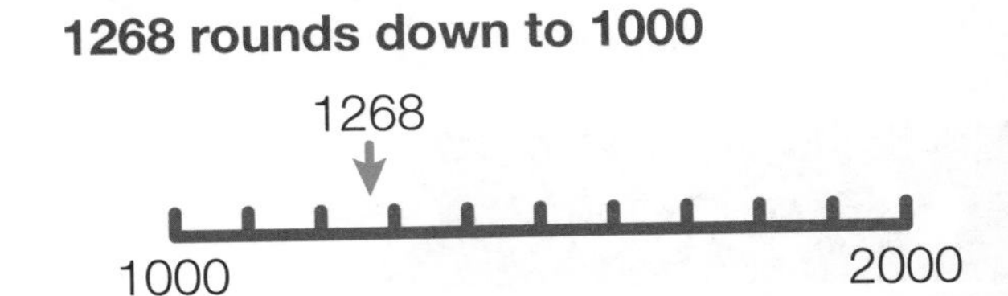

Revise

It isn't hard, you just need to think about where they are on the number line.

We often round numbers to the nearest power of 10 (that's 10, 100, 1000, 10,000, 100,000, and so on).

Rounding to the nearest 10,000: 4235 rounds down to zero and 6249 rounds up to 10,000.

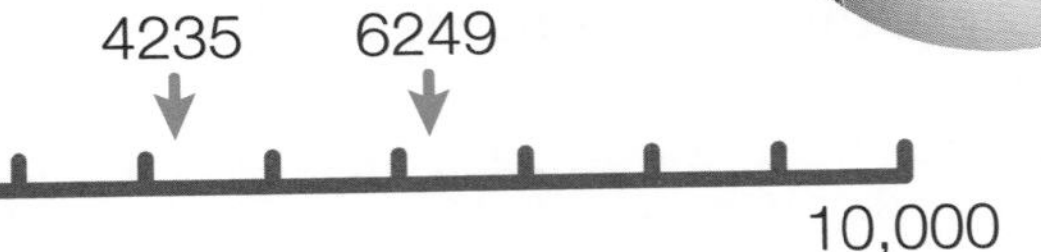

Rounding to the nearest 100,000: 344,235 rounds down to 300,000 and 689,249 rounds up to 700,000.

344,235 689,249

0 500,000 1,000,000

Tips

- Always think carefully about what you want to round to: nearest 10, 100, 1000 and so on, and then think about the part of the number line the number is on. So:
  635,850 rounds to the nearest 10 as 635,850
  635,850 rounds to the nearest 100 as 635,900
  635,850 rounds to the nearest 1000 as 636,000
  635,850 rounds to the nearest 10,000 as 640,000
  635,850 rounds to the nearest 100,000 as 600,000

Talk maths

Write six different numbers between zero and one million. Read aloud each number to a friend and challenge them to round it to each power of 10 from 10 to 10,000.

347,248 rounded to the nearest 1000 is 347,000.
What is 54,250 rounded to the nearest 100?

✓ Check

Complete the table below.

	Rounded to nearest 10	Rounded to nearest 100	Rounded to nearest 1000	Rounded to nearest 10,000	Rounded to nearest 100,000
67					
145					
3320					
78,249					
381,082					
555,555					

Problems

Brain-teaser 54,527 people watch a football match.
What is this rounded to the nearest 10,000? ____________

Brain-buster A famous footballer normally gets paid £346,000 per match! If he scores a goal his pay is rounded up to the next 100,000. If he doesn't score a goal it is rounded down to the nearest 100,000. How much does he lose if he doesn't score, and how much does he gain if he does?

Loss if fails to score: £____________ ; Gain if scores: £____________

Roman numerals

What are the rules for making 4 and 9 with Roman numerals?

Recap

Look at the clock and check that you know the Roman numerals for numbers 1–12.

Revise

Number	1	2	3	4	5	6	7	8	9	10
Roman numeral	I	II	III	IV	V	VI	VII	VIII	IX	X
Number	11	12	13	14	15	16	17	18	19	20
Roman numeral	XI	XII	XIII	XIV	XV	XVI	XVII	XVIII	IXX	XX
Number	30	40	50	60	70	80	90	100	500	1000
Roman numeral	XXX	XL	L	LX	LXX	LXXX	XC	C	D	M

With a bit of brain power, you can use this chart to find out any Roman numeral up to 1000!

Check

1. Write these Roman numerals in numbers.

   a. VIII 8 b. XXIII 23 c. CCC 300 d. XCV ______

   e. CIV 104 f. CXL 130 g. DCX 420 h. CM ______

2. Write these numbers in their Roman numeral equivalent.

   a. 22 ______ b. 41 ______ c. 55 ______ d. 93 ______

   e. 112 ______ f. 160 ______ g. 212 ______ h. 965 ______

Problems

Brain-buster The Romans left Britain in the year AD410, 465 years after they first arrived. Use Roman numerals to write the date they left, and the number of years they spent in Britain.

Date: ______ Years in Britain: ______

Mental methods for adding and subtracting

Recap

You will probably know several ways of doing mental calculations.

You must know your number bonds: 7 + 8 = 15   15 – 8 = 7   15 – 7 = 8

Partitioning numbers is important too: **25** + **12** = **37**

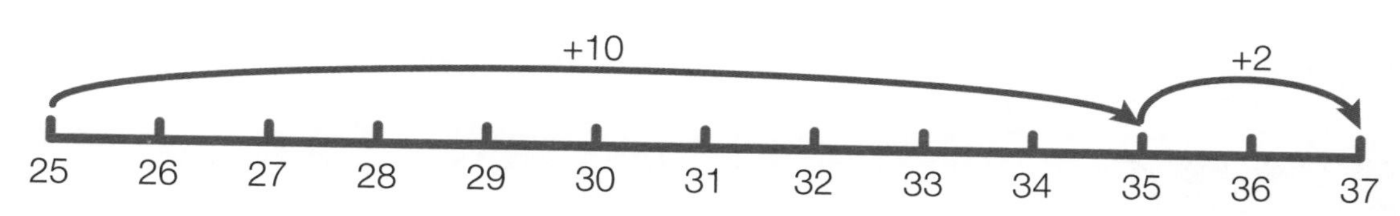

Revise

Remember, adding 99 is easy: add 100 and take away 1!

Mental methods can work just as well for larger numbers, but you need to be confident and know your limits!

45,356 + 12,103 ✓ **There is no carrying necessary; just add each column.**

123,729 + 943,509 ✗ **Too much carrying!**

34,302 – 8753 ✗ **Too much carrying!**

16,583 – 8000 ✓ **The 100s, 10s and 1s stay exactly the same!**

✓ Check

1. Add these numbers using mental methods.

a. 46 + 50 = _____   **b.** 127 + 99 = _____

c. 3274 + 2002 = _____   **d.** 2500 + 7454 = _____

e. 120,000 + 10,320 = _____

2. Subtract these numbers using mental methods.

a. 80 – 46 = _______   **b.** 160 – 65 = _____

c. 345 – 99 = _______   **d.** 4000 – 2500 = _____

e. 275,675 – 10,000 = _____

Problems

Brain-teaser Jason has read 123 pages of his book. If he reads another 150 pages he will finish it. How many pages does the book have altogether?

____

Brain-buster Armchairs cost £299 and sofas cost £499. How much would two armchairs and one sofa cost altogether?

____

Adding large numbers

We can arrange numbers in their place-value columns.

Recap

There are formal written methods for adding numbers. You may have been taught methods a bit different to this one. You should use whichever method you are comfortable with – as long as you get the right answer!

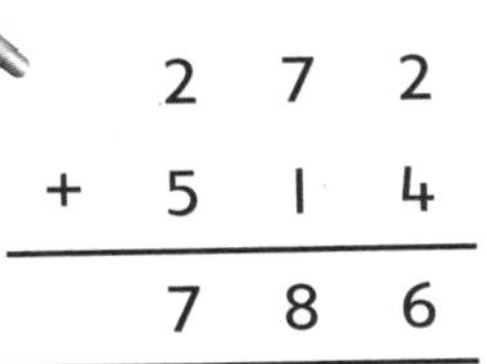

Revise

We know that the place-value columns continue for 1000s, 10,000s, 100,000s and so on.

We can use formal written methods using these columns.

	1,000,000s	10,000s	1000s	100s	10s	1s
	6	6	4	5	7	2
+	1	5	3	0	5	4
	8	1	7	6	2	6
	1			1		

Don't forget, you can add as many numbers as you like in the columns!

Tips

If in doubt, work it out!

- Lay out your work neatly, showing the + sign and carefully exchanging the numbers between columns.
- When adding two or more numbers some people prefer to write the larger number in the top row of the calculation, but it really doesn't matter which way round you arrange them – the answer will always be the same!

	4	1	8	2	1	6
+	3	2	5	2	7	4
	7	4	3	4	9	0
		1			1	

Talk maths

Look at the addition below and talk it through aloud, explaining how each stage was done. Make sure you work in the correct order.

	2	3	7	1	6	2
+	4	8	4	7	5	3
	7	2	1	9	1	5
	1	1		1		

Use estimation to quickly see if your answers are about right.

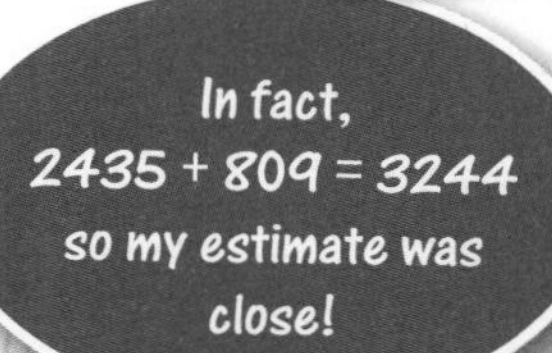

2435 + 809 is around 2500 + 1000, so the answer will be around 3500.
In fact, 2435 + 809 = 3244 so my estimate was close!

✓ Check

1. Complete each of these additions.

a.

	2	4	3	5
+		8	0	9

b.

	7	4	3	2
+	4	8	7	7

c.

	2	4	3	5	7
+	4	5	8	2	3

d.

	2	4	5	0	2	0
+	3	7	6	2	0	9

2. Use squared paper to write and complete each of these additions.

a. 2459 + 3507

b. 23,417 + 46,219

c. 124,467 + 89,458

d. 231,472 + 238,419 + 121,615 + 67,424

⚠ Problems

Brain-teaser This chart shows the populations of three imaginary cities.

City	Bim	Bam	Bom
Population	236,325	143,544	367,269

Are the combined populations of Bim and Bam larger than the population of Bom? ____________

Brain-buster If the population of each city increased by 50,000 people, would the total population of the three cities be more than one million people? ____________

Subtracting large numbers

Recap

There are formal written methods for subtracting numbers. You may have been taught methods a bit different to this one. You should use whichever method you are comfortable with – as long as you get the right answers!

	$^{3}\cancel{4}$	$^{1}1$	5
−	2	3	4
	1	8	1

Notice how
we exchange one
100 for ten 10s.

Revise

Just like with addition, we can use the place-value columns to help us subtract larger numbers.

	1,000,000s	10,000s	1000s	100s	10s	1s
	$^{2}\cancel{3}$	$^{13}\cancel{4}$	$^{10}\cancel{1}$	$^{1}2$	4	6
−	1	6	5	3	0	4
	1	7	5	9	4	2

You need to be very careful at each stage of a written subtraction. Look at this one:

	$^{5}\cancel{6}$	$^{12}\cancel{3}$	$^{9}\cancel{0}$	$^{1}2$	$^{3}\cancel{4}$	$^{1}5$
−	2	3	5	4	2	9
	3	9	4	8	1	6

Look at what
you must do if you want
to exchange ten of one number
for a larger number but
the next column
has a zero.

Tips

Here are some
useful subtraction
hints.

- Remember, you can check your subtractions by adding your answer to the number you took away.
  243 − 175 = 68 checking… **68 + 175 = 243** correct!
- If you have a method you like, stick to it, practise it, and always check your answers.

Talk maths

Look at the subtraction below and talk it through aloud, explaining how each stage was done. Make sure you work in the correct order.

	$^{3}\not{4}$	$^{1}3$	5	5	$^{5}\not{6}$	$^{1}2$
−	2	4	5	4	2	6
	1	9	0	1	3	6

Use estimation to quickly see if your answers are *about right*.

1405 – 950 is around 1400 – 900, so the answer will be around 500.
In fact, 1405 – 950 = 455, so my estimate was close!

✓ Check

1. Complete each of these subtractions.

a.

	3	7	4
−	2	3	5

b.

	7	4	2	8
−	3	2	6	5

c.

	4	3	2	6	2	5
−	2	4	3	2	0	6

2. Use squared paper to write out and complete each of these subtractions.

a. 235 – 116 **b.** 4823 – 2550 **c.** 13,274 – 9306

d. 10,206 – 6345 **e.** 240,231 – 123,308

⚠ Problems

Brain-teaser This chart shows the populations of three imaginary cities.

City	Bim	Bam	Bom
Population	236,325	143,544	367,269

Which is bigger, the difference between the populations of Bim and Bam, or the difference between the populations of Bim and Bom? ______________________

Brain-buster There is a fourth city called Bem. It has a population that is 154,289 less than Bom. What is the population of Bem? ______________________

Multiples and factors

Factors are the numbers that we multiply together to get multiples.

Recap

A **multiple** is a number that is made by multiplying two numbers.

4 × 3 = 12

12 is a **multiple** of 3, and it is also a **multiple** of 4.
We can also say that 3 and 4 are **factors** of 12.

Revise

Factors are easiest to find as pairs.

12 = (1 × 12), (2 × 6) or (3 × 4)

So the factors of 12 are 1, 2, 3, 4, 6 and 12.

15 = (1 × 15) or (3 × 5)

So the factors of 15 are 1, 3, 5 and 15.

Sometimes two different numbers will share the same factor.
We call this a **common factor**.

3 is a common factor of 12 and 15.

Remember, we can also say that 12 is a multiple of 1, 2, 3, 4 and 6.

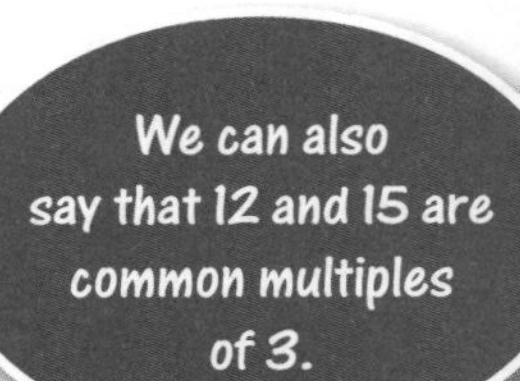

Here's some help with multiples and factors.

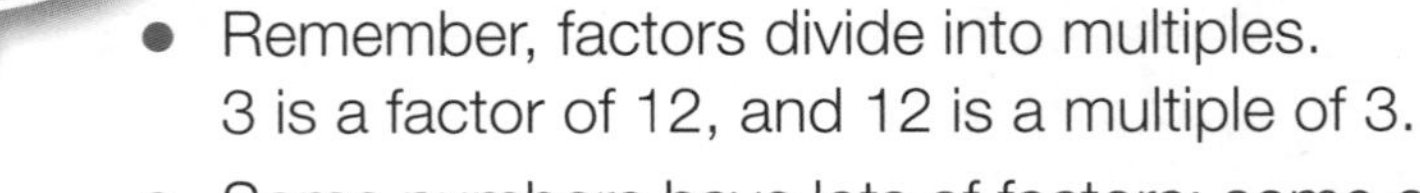

- Remember, factors divide into multiples.
  3 is a factor of 12, and 12 is a multiple of 3.
- Some numbers have lots of factors; some only have 2.
- All prime numbers only have themselves and 1 as factors.
- A factor that is *also* a prime number is called, wait for it, a *prime factor*!

Talk maths

Close this book, and then explain to an adult what a factor is and what a multiple is. Use examples to help you.

Then open the book and check how you did.

DID YOU KNOW?

Did you know that multiples go on forever? Don't try it out, just trust me!

✓ Check

1. **Write down all the factors of 6.** ____________________
2. **Write down five multiples of 4.** ____________________
3. **Write all the factor pairs for each of these numbers.**
   - **a.** 15 ____________________ **b.** 27 ____________________
   - **c.** 24 ____________________ **d.** 30 ____________________
4. **Find the common factors of these numbers.**
   - **a.** 12 and 16 ____________________ **b.** 15 and 20 ____________________
   - **c.** 28 and 40 ____________________ **d.** 50 and 100 ____________________

Problems

Brain-teaser Selina has 24 small chocolates and she wants to share them equally with some of her friends. Complete the chart to show how many children she **could** share them equally between, and how many chocolates each person would get.

Children	1	2						
Chocolates	24	12						

What would happen if Selina tried to share her chocolates between five friends?

____

Brain-buster Everyone knows there are 365 days in a year, and 7 days in a week.

Are there exactly 52 weeks in a year? ____________________

Explain your answer. ____________________

Prime numbers

1 is not counted as a prime number.

Recap

A number that can only be divided by itself or 1, with no remainder, is called a prime number. For example, 2, 3 and 5 are all prime numbers.

2 is the only even prime number. All other even numbers can be divided by 2 as well as 1 and themselves.

Revise

~~1~~	(2)	(3)	~~4~~	(5)	~~6~~	(7)	~~8~~	~~9~~	~~10~~
(11)	~~12~~	(13)	~~14~~	~~15~~	~~16~~	(17)	~~18~~	(19)	~~20~~
~~21~~	~~22~~	(23)	~~24~~	~~25~~	~~26~~	~~27~~	~~28~~	(29)	~~30~~
(31)	~~32~~	~~33~~	~~34~~	~~35~~	~~36~~	(37)	~~38~~	~~39~~	~~40~~
(41)	~~42~~	(43)	~~44~~	~~45~~	~~46~~	(47)	~~48~~	~~49~~	~~50~~
~~51~~	~~52~~	(53)	~~54~~	~~55~~	~~56~~	~~57~~	~~58~~	(59)	~~60~~
(61)	~~62~~	~~63~~	~~64~~	~~65~~	~~66~~	(67)	~~68~~	~~69~~	~~70~~
(71)	~~72~~	(73)	~~74~~	~~75~~	~~76~~	~~77~~	~~78~~	(79)	~~80~~
~~81~~	~~82~~	(83)	~~84~~	~~85~~	~~86~~	~~87~~	~~88~~	(89)	~~90~~
~~91~~	~~92~~	~~93~~	~~94~~	~~95~~	~~96~~	(97)	~~98~~	~~99~~	~~100~~

Look at the numbers 1 to 10. We can circle 2 as a prime number. We know that all even numbers can be divided by 2, so we can delete all other even numbers because we know that none of these can be prime numbers.

We can also circle 3, and then delete 9. We know from the times tables that 9 can be divided by 3, so it cannot be a prime number.

I have circled all the prime numbers for you.

- There are rules that can help you decide if any number is a prime or not.
- A number that is even can be divided by 2, so no even numbers are prime numbers, except 2 itself of course!
- Add the digits of the number together. If the sum can be divided by 3, so can the number, and so it is not a prime number, for example 207: 2 + 0 + 7 = 9, 9 can be divided by 3, so 207 is not a prime number!
- If a number ends in 0 or 5 it can be divided by 5, so it is not a prime number, for example 115 is not a prime number.

Warning! These rules only *help* you to decide, you may still need to check for other prime factors!

Play *Prime Time!*

Talk maths

97 5 2 11 41

Challenge an adult or a friend to a game of *Prime Time*. You need something to time minutes on, such as a stopwatch. You will also need a pencil and paper, for keeping scores and remembering which numbers have been used.

Take it in turns to say a number and challenge the other player to decide whether it is a prime number or not, and record how longit took to answer the question. If challenged, a player must prove why their answer is yes or no, with a sensible explanation.

DID YOU KNOW?

Mathematicians are still discovering new prime numbers. Imagine how enormous those numbers must be!

✓ Check

1. **What is a prime number?** ______________________________

2. **Write all the prime numbers between 1 and 20 (there are eight altogether).**

____

3. **Say which of these numbers are prime, and explain each of your answers.**

   a. 25 ______________________________

   b. 71 ______________________________

   c. 87 ______________________________

4. **Can you think of a prime number greater than 100?** ______________________________

⚠ Problems

Brain-teaser 77 cannot be divided by 2, 3 or 5. Does this make it a prime number? Explain your answer.

____

Brain-buster Mohammed says that 7 is a prime number, and so is 17, so 27 must also be a prime number. Explain why he is wrong.

____

Multiplying large numbers

Recap

Remember, the numbers are arranged in their place-value columns.

There are formal written methods for multiplying numbers. You may have been taught methods a bit different to this one. You should use whichever method you are comfortable with – as long as you get the right answers!

	6	3				3	2	5				4	6	2	5
×		7			×			6			×				5
4	4	1			1	9	5	0			2	3	1	2	5
	2					1	3					3	1	2	

Revise

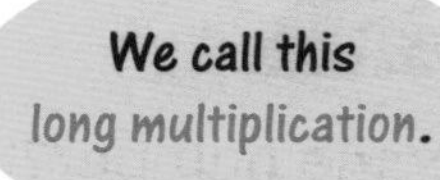
We call this long multiplication.

We know that the place-value columns continue for thousands, ten-thousands, hundred-thousands and so on. We can use formal written methods using these columns.

When multiplying two numbers larger than 10, multiply each digit on the top by each digit on the bottom, carrying numbers to the next column along, when necessary.

You can multiply first by the 10s, or the 1s; the answer will still be the same.

	4	6	
×	1	5	
2	3^3	0	
4	6	0	+
6	9	0	

Answer: 690

Tips

- Most people find it easier to put the larger number on the top, although it doesn't really matter which way round you arrange them – the answer will still be the same.
- Remember, the 1 digit in 14 is 10.
- It doesn't matter if you start with the 10s or the 1s.

		3	2	
	×	1	4	
(32 × 4)	1	2	8	
(32 × 10)	3	2	0	+
	4	4	8	

OR

		3	2	
	×	1	4	
(32 × 10)	3	2	0	
(32 × 4)	1	2	8	+
	4	4	8	

Remember that zeros still have to be multiplied, and anything times zero is zero!

Talk maths

Look at the long multiplications below and talk them through aloud, saying how each stage was done.

	3	2	5			2	7				4	3					4	3	6	
×			3		×	2	1				×	3	5			×		2	1	
	9	7	5			2	7				2	1^{1}	5				4	3	6	
		1			5^{1}	4	0	+		1	2	9	0	+		8	7^{1}	2	0	+
					5	6	7			1	5	0	5			9	1	5	6	
											1					1				

1. Complete each of these long multiplications.

a.

	2	1
×	1	3

b.

	2	3
×	2	4

c.

		4	5
	×	3	1

d.

		6	3
	×	5	6

2. On squared paper, write out and complete each of these long multiplications.

a. 15 × 21 **b.** 26 × 32 **c.** 53 × 15 **d.** 33 × 40

Problems

Brain-teaser A school's tuck shop sells cartons of fruit juice. Each carton cost 15p.

If 43 cartons are sold, how much money will be collected? ______________________

Brain-buster There are 475 children in a school. The school is raising money for charity with a sponsored walk, and they hope to raise £6000. If each child raises £13 will they hit their target? Explain your answer.

____

____

Dividing large numbers

Recap

There are formal written methods for dividing numbers. You may have been taught methods a bit different to this one. You should use whichever method you are comfortable with – as long as you get the right answers!

		1	2				2	4	
	3	3	6			4	9	$^{1}6$	

For 36 ÷ 3 = 12 we say 36 divided *by* 3 equals 12.

Revise

In short division we move forward remainders. Sometimes there is a remainder at the end.

		1	2	r1					2	4	r3	
	3	3	7					4	9	$^{1}9$		

There is also a method called long division. You don't have to use that method yet, but if you know how to use it no one is going to stop you!

We can still use short division when we are dividing by 2-digit numbers. We just need to look for the first whole number that the 2-digit number can divide into, the rest is the same.

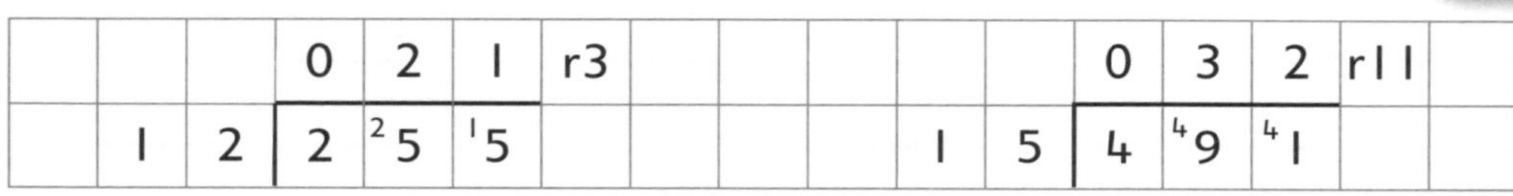

			0	2	1	r3						0	3	2	r11	
	1	2	2	$^{2}5$	$^{1}5$					1	5	4	$^{4}9$	$^{4}1$		

- Lay out your work carefully. Use squared paper to help you.
- You can always check your answer by multiplying the answer by the number you divided by, and then add the remainder.

Talk maths

Remember that zero divided by anything is zero.

Look at this division and talk it through aloud, saying how each stage was done.

	1	2	4	r3
4	4	9	19	

✔ Check

1. Complete each of these divisions.

a.

5	9	5

b.

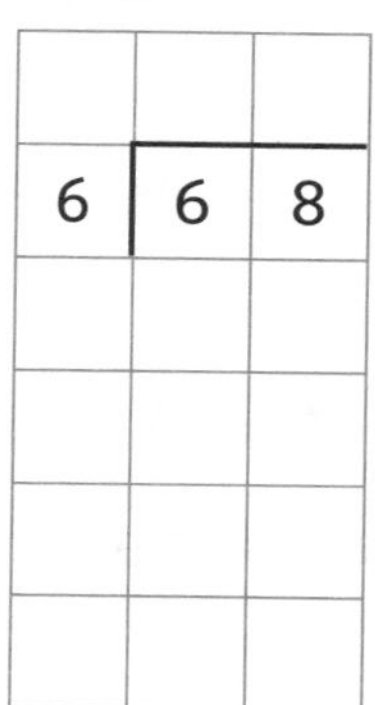

6	6	8

c.

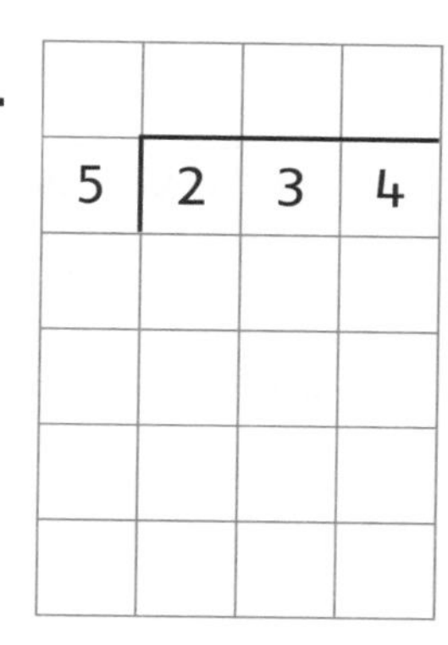

5	2	3	4

d.

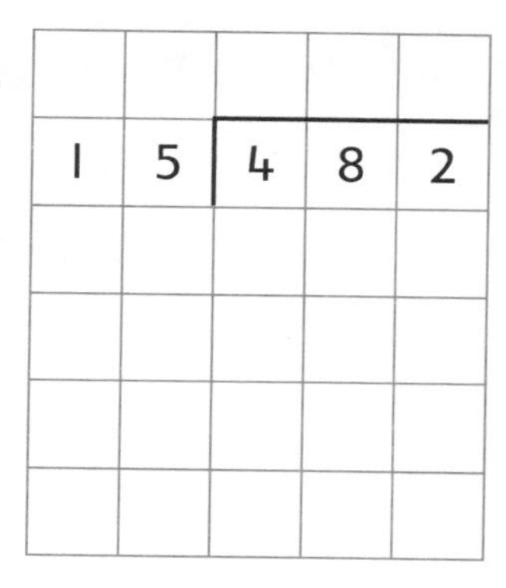

1	5	4	8	2

2. On squared paper, write out and complete each of these divisions.

a. $98 \div 7$ b. $125 \div 5$ c. $522 \div 8$ d. $318 \div 15$

⚠ Problems

Brain-teaser There are 500 staples in a box. If the teacher's staple gun holds 40 staples, how many times can she refill her staple gun? ____________

Brain-buster Mako makes a large bowl of popcorn for her class party. If there are 246 pieces of popcorn, and there are 21 children in the class, how many pieces of popcorn will each child get if it is divided equally?

____

Mako's teacher eats any remaining pieces. How many pieces does he get? ____________

Mental methods for multiplying and dividing

Recap

You know how to use a multiplication square.

Multiplication squares help to show us that division is the *inverse* of multiplication.

So, we can say

6 × 7 = 42 **7 × 6 = 42**

42 ÷ 6 = 7 **42 ÷ 7 = 6**

×	1	2	3	4	5	6	7	8	9	10
1	1	2	3	4	5	6	7	8	9	10
2	2	4	6	8	10	12	14	16	18	20
3	3	6	9	12	15	18	21	24	27	30
4	4	8	12	16	20	24	28	32	36	40
5	5	10	15	20	25	30	35	40	45	50
6	6	12	18	24	30	36	42	48	54	60
7	7	14	21	28	35	42	49	56	63	70
8	8	16	24	32	40	48	56	64	72	80
9	9	18	27	36	45	54	63	72	81	90
10	10	20	30	40	50	60	70	80	90	100

Revise

You can use your times tables to help you with harder mental calculations.

4 × 30 = 120 (We know that 4 × 3 = 12, so **4 × 30 = 120**)

520 × 6 = 3120 (We know that 5 × 6 = 30, so **500 × 6 = 3000**, and **20 × 6 = 120**)

123 ÷ 3 = 41 (We know that 12 ÷ 3 = 4, so **120 ÷ 3 = 40**, and **3 ÷ 3 = 1**)

2816 ÷ 4 = 704 (We know that 28 ÷ 4 = 7, so **2800 ÷ 4 = 700**, and **16 ÷ 4 = 4**)

Too tricky? If in doubt, write it out!

Check

1. **Solve these multiplications mentally.**

   a. 4 × 50 = _______ **b.** 320 × 3 = _______ **c.** 2 × 4444 = _______

   d. 5 × 6000 = _______ **e.** 7 × 2050 = _______

2. **Now try these divisions using mental methods.**

   a. 300 ÷ 6 = _______ **b.** 129 ÷ 3 = _______ **c.** 5005 ÷ 5 = _______

   d. 3608 ÷ 4 = _______ **e.** 2828 ÷ 7 = _______

Problems

Brain-buster Six people share a lottery ticket that wins £12,300. If they share the winning amount equally, how much will they each receive? _______________

Square numbers and cube numbers

Recap

A square number is a number multiplied by itself, for example, 2 squared is **2 × 2 = 4**

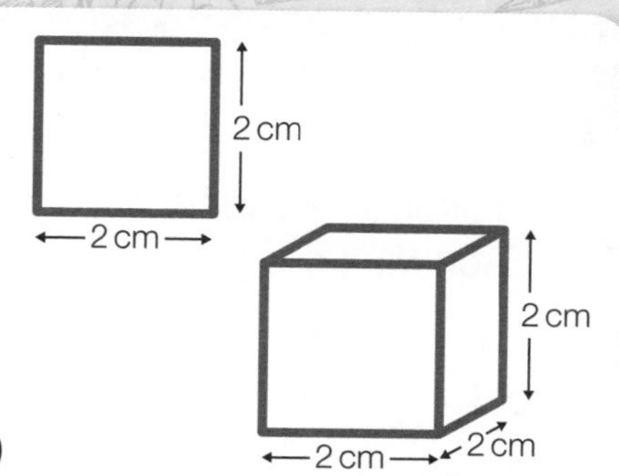

A cube number is a number multiplied by itself, and then by itself again, for example,

2 cubed is **2 × 2 × 2 = 8** (2 × 2 = 4, then 4 × 2 = 8)

Revise

We use special numbers, called **powers**, to show square and cube numbers.

For 5 squared, instead of **5 × 5** we say **5^2 = 25**

For 4 cubed, instead of **4 × 4 × 4** we say **4^3 = 64**

Five squared equals twenty-five.
Four cubed equals sixty-four.

✔ Check

Complete this chart, then use it to practise and learn square and cube numbers.

1	2	3	4	5	6	7	8	9	10
1^2	2^2								
1 × 1	2 × 2								
1	4								
1^3	2^3								
1 × 1 × 1	2 × 2 × 2								
1	8								

Problems

Brain-teaser Some children organise a five-a-side football match. Each player in Sanjay's team scored five goals. How many goals did his team score altogether? ____________

Brain-buster A farmer stores apples in boxes. Each box holds nine layers of apples; each layer is nine apples wide and nine apples long. How many apples are there in each box?

____

Multiplying and dividing by 10, 100 and 1000

Please don't say *just add a zero.* That doesn't work for decimals!

Recap

Our number system is arranged in **powers of 10.** When we multiply a number by 10 we make each digit 10 times bigger. Each digit moves one place to the left.

6 × 10 = 60　43 × 10 = 430　257 × 10 = 2570
0.4 × 10 = 4　0.07 × 10 = 0.7

When we divide a number by 10, we make each digit 10 times smaller. Each digit moves one place to the right.

6 ÷ 10 = 0.6　43 ÷ 10 = 4.3　257 ÷ 10 = 25.7
0.4 ÷ 10 = 0.04　0.07 ÷ 10 = 0.007

The decimal point doesn't move! It is always between the ones and tenths.

Revise

Operation	Fact	Example 1	Example 2
× 10	Digits move one place left	65 × 10 = 650	7 × 10 = 70
× 100	Digits move two places left	65 × 100 = 6500	7 × 100 = 700
× 1000	Digits move three places left	65 × 1000 = 65,000	7 × 1000 = 7000
÷ 10	Digits move one place right	65 ÷ 10 = 6.5	7 ÷ 10 = 0.7
÷ 100	Digits move two places right	65 ÷ 100 = 0.65	7 ÷ 100 = 0.07
÷ 1000	Digits move three places right	65 ÷ 1000 = 0.065	7 ÷ 1000 = 0.007

Tips

- Think about the place-value columns:

1,000,000s	100,000s	10,000s	1000s	100s	10s	1s	0.1s	0.01s	0.001s
				2	5	7 •			

- For any calculation, think about the number becoming bigger or smaller, moving it to the left or the right. Try this for **257 × 1000**, then for **257 ÷ 1000**. Then try some other numbers.

Talk maths

Practise using the correct vocabulary.

23 times 1000 is 23,000. It is now 1000 times bigger.

0.06 divided by 10 is 0.006. It is now 10 times smaller.

3 divided by 100 is 0.03. It is now 100 times smaller.

0.023 times 100 is 2.3. It now is 100 times bigger.

Now, using a pencil and paper, explain to an adult or a friend how to multiply and divide by 10, 100 and 1000.

✔ Check

1. Complete these grids.

		× 10	× 100	× 1000
	3	30	300	3000
÷ 10	0.3	3		
÷ 100	0.03		3	
÷ 1000	0.003			3

		× 10	× 100	× 1000
	27			
÷ 10				
÷ 100				
÷ 1000				

		× 10	× 100	× 1000
	48			
÷ 10				
÷ 100				
÷ 1000				

		× 10	× 100	× 1000
	317			
÷ 10				
÷ 100				
÷ 1000				

⚠ Problems

Brain-teaser Tim says that an aeroplane flying over his house is 1000 times higher than the roof of his house. His house is 32 feet tall. How high up is the aeroplane?

____

Brain-buster A new-born piglet weighs just one hundredth of its mother's weight. If its mother weighs 135.6kg, what weight will the piglet be? ________________

Scaling and rates

Recap

We can easily calculate fractions of quantities.

A farmer sells half his flock of 24 sheep.
$\frac{1}{2}$ of 24 sheep = 12 sheep.

Calculate **scale**: The building is 100 times bigger than the model.

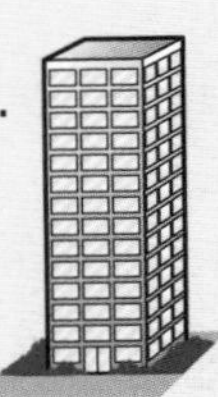

The model is 20cm tall, so the building must be 2000cm, or 20 metres tall.

We can also calculate the rate that things happen.

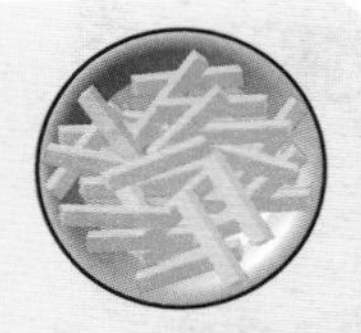

If I eat 25 chips in five minutes, I have eaten five chips per minute.

Revise

Here are some problems involving scales and rates.

Examples of scales:

A child is half the height of his dad. If the child is 90cm, the father must be 180cm.

A car is four times the size of a model. If the car is 5 metres long the model must be 1.25 metres.

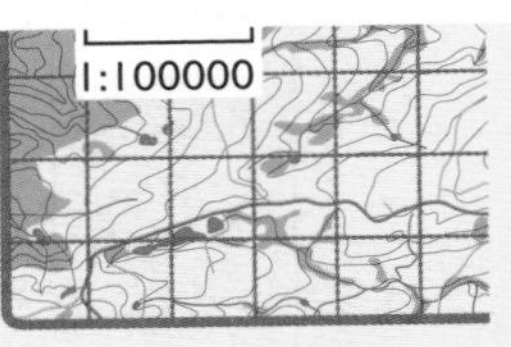

The scale of a map is 1:100,000. This means that every centimetre on the map equals a kilometre in real life.

Examples of rates:

In a traffic survey, 12 cars drive past a school in one hour. We can estimate that in six hours 72 cars will go past the school (6 × 12 = 72 cars).

A bathtub fills up at a rate of 10 litres per minute. If it takes 15 minutes to fill the bath, the bath must have a capacity of 150 litres (10 × 15 = 150 litres).

Ten people per minute go into a cinema. If it takes 40 minutes to fill all the seats, the cinema must hold 400 people (10 × 40 = 400 people).

Tips

- When calculating scales, remember you can use the inverse too.
- If a road is 12km long then we know it will be 12cm on a map with a scale of 1 to 100,000.
- If a 10cm model is one fifth the size of an object, the object will be 50cm high.

Talk maths

Look at this map and check that you understand the scale. Talk about it with someone, discussing what the real-life distances will be between different features on it.

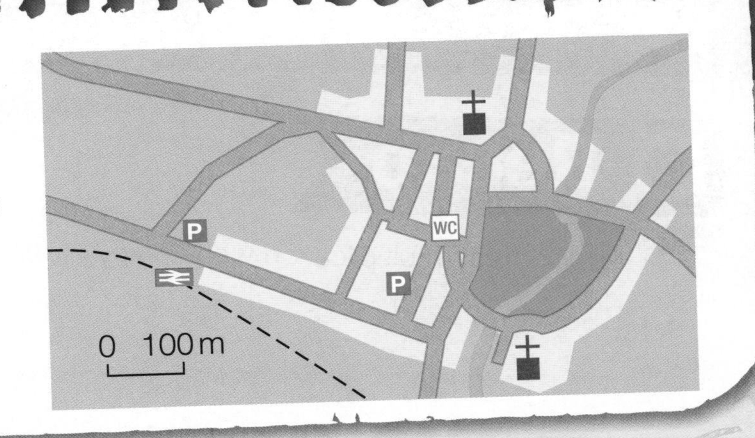

✓ Check

1. **Calculate these fractions.**

   a. $\frac{1}{2}$ of 6 cakes = ___________

   b. $\frac{1}{4}$ of 20 adults = ___________

   c. $\frac{1}{3}$ of 66 animals = ___________

   d. $\frac{3}{4}$ of 100 children = ___________

2. **Draw two rectangles half the length and half the width of those shown.**

3. **A teacher makes a scale model of her classroom. She builds everything at a scale of 1 to 20. Complete the chart to show the sizes of each item in her model.**

Item	Room	Table	Chair	Cupboard	Waste basket
Real height	280cm	90cm	40cm	170cm	25cm
Model height					

⚠ Problems

Brain-teaser Mason measures his heart rate. It is 60 beats per minute.

How many times will his heart beat in an hour? ___________

How many times in a day? ___________

Brain-buster A sculptor makes a 12cm model of an athlete. The athlete is 180cm tall.

What scale is the model? (Give your answer as a fraction.) ___________

Using all four operations

Recap

Addition and multiplication work in any order, division and subtraction do not.

$3 + 4 = 4 + 3$ $5 \times 6 = 6 \times 5$ $12 - 3 \neq 3 - 12$ $10 \div 2 \neq 2 \div 10$

Revise

Calculations and problems involving more than one operation are called **multi-step**. You must do one calculation at a time.

And you must do them in the right order!

The right order is division and multiplication first, followed by addition and subtraction, working from left to right.

Look at this calculation: 16 ÷ 4 + 2 × 5 − 12

Division first (**16 ÷ 4 = 4**)	**4 + 2 × 5 − 12**
Multiplication next (**2 × 5 = 10**)	**4 + 10 − 12**
Then addition (**4 + 10 = 14**)	**14 − 12**
And last subtraction (**14 − 12 = 2**)	**Answer** = 2

Now let's try this one: 24 − 3 × 5 + 10 ÷ 2

Multiplication first (**3 × 5 = 15**)	**24 − 15 + 10 ÷ 2**
Division next (**10 ÷ 2 = 5**)	**24 − 15 + 5**
Then subtraction (**24 − 15 = 9**)	**9 + 5**
And last addition (**9 + 5 = 14**)	**Answer** = 14

Tips

OK, everything seems to be in order here, so let's think about estimation.

- When you use maths to solve problems, think about the operations you will use, and start with an estimate. A quick estimate for $12 \times 6 - 24 \div 3 + 15$ would be to do the division and multiplication $72 - 8 + 15$ then round to get $70 - 10 + 15$, giving an estimate of 75 (the answer is 79).

Talk maths

Look at these problems, which need more than one calculation. Talk about each problem with an adult or a friend, writing down the calculation and explaining why you need each operation. Then use your mental maths skills to estimate the answer before checking.

How much would one baguette, 2 meal deals and 4 drinks cost altogether?

Baguettes	£4
Pizzas	£6
Meal deal (pizza, chips, salad)	£8
Drinks	£2

How much change would you get from £100 if you bought 4 meal deals, 3 baguettes, 5 pizzas and 12 drinks?

Now make up some problems of your own.

✓ Check

1. Complete these calculations.

a. $6 \times 2 - 3 =$ ________

b. $6 + 2 \times 3 =$ ________

c. $6 \div 2 + 3 =$ ________

d. $6 - 2 \times 3 =$ ________

2. Mark each of these calculations right (✓) or wrong (✗).

a. $3 \times 8 + 16 \div 4 - 12 \div 6 = 26$ ____

b. $20 - 6 \times 2 = 28$ ____

c. $13 - 25 \div 5 \times 2 = 3$ ____

d. $12 - 6 + 3 = 3$ ____

e. $45 \div 5 + 4 \times 5 - 3 \times 7 = 8$ ____

f. $70 - 20 \times 3 + 50 \div 10 = 20$ ____

3. Add the missing signs to make each of these true.

a. $5 \times 5 + 12$ ____ $2 - 3 \times 4 = 19$

b. 12 ____ 3 ____ $14 \div 7 + 2 \times 4 = 10$

⚠ Problems

Brain-teaser A customer asks for two cabbages and five onions.

How much will this cost altogether? ______________

Brain-buster A customer requests three cabbages, four onions, one lettuce and a cucumber. How much change will she receive from a £10 note?

____

Grocer's price list

	Cabbages	80p
	Cucumbers	£1.20
	Lettuce	90p
	Onions	30p

Comparing and ordering fractions

Recap

Fractions have a numerator and a denominator.
They show us proportions of a whole.
One out of four pieces of this pizza has been eaten.

numerator — $\frac{1}{4}$ — **denominator**

We can say $\frac{1}{4}$ has been eaten…

… and $\frac{3}{4}$ has not been eaten!

Revise

We can compare and order fractions by giving them the same denominators.
To do this we must understand **equivalent fractions.**
This pizza has been cut into eight equal pieces, or eighths.

$\frac{2}{8} = \frac{1}{4}$

These fractions are **equivalent** because they are **the same proportion of the whole**.
We can check this by changing either one of them:

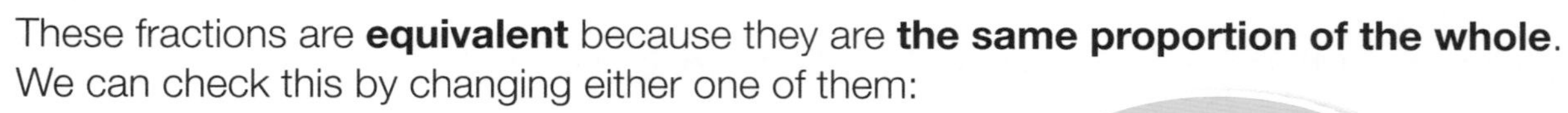

$\frac{2 \div 2}{8 \div 2} = \frac{1}{4}$ $\frac{1 \times 2}{4 \times 2} = \frac{2}{8}$

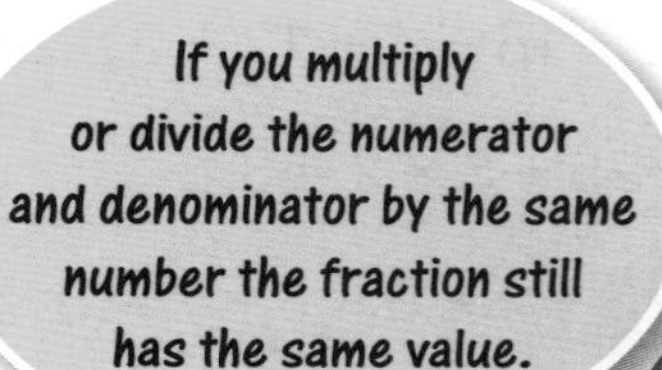

If you multiply or divide the numerator and denominator by the same number the fraction still has the same value.

To compare and order fractions, it is easier if they all have the same denominator.

Which is bigger, $\frac{1}{3}$ or $\frac{2}{5}$? We can change both fractions into fifteenths and then compare:

$\frac{1 \times 5}{3 \times 5} = \frac{5}{15}$ $\frac{2 \times 3}{5 \times 3} = \frac{6}{15}$

So, $\frac{2}{5}$ is bigger than $\frac{1}{3}$. 15 is the **common denominator**.

Tips

- Finding a common denominator is the same as finding a lowest common multiple.
- Remember:
  < means **is smaller than**
  > means **is bigger than**
  We can say $\frac{2}{5} > \frac{1}{3}$ and $\frac{1}{3} < \frac{2}{5}$

Talk maths

Look at the fractions in the circle and make statements using < and >.

Use these phrases to help you.

The lowest common denominator of ___ and ___ is ___

___ is bigger than ___

___ is smaller than ___

___ is equivalent to ___

Can you make a statement that includes three fractions?

$\frac{1}{3}$ $\frac{1}{4}$ $\frac{1}{2}$ $\frac{3}{5}$ $\frac{1}{5}$ $\frac{3}{4}$ $\frac{2}{3}$ $\frac{4}{5}$ $\frac{5}{6}$ $\frac{2}{6}$ $\frac{5}{7}$ $\frac{4}{8}$ $\frac{1}{7}$ $\frac{1}{8}$ $\frac{7}{8}$ $\frac{6}{8}$ $\frac{8}{9}$ $\frac{2}{9}$ $\frac{6}{9}$ $\frac{3}{9}$

✓ Check

1. Change each fraction to give it a denominator of 8.

a. $\frac{1}{2}$ = ______ **b.** $\frac{1}{4}$ = ______ **c.** $\frac{3}{4}$ = ______ **d.** 1 whole = ______

2. Change each fraction to give it a denominator of 12.

a. $\frac{1}{2}$ = ______ **b.** $\frac{1}{4}$ = ______ **c.** $\frac{2}{3}$ = ______ **d.** $\frac{5}{6}$ = ______

3. True or false? Circle the correct statements.

a. $\frac{1}{3} = \frac{2}{6}$ **b.** $\frac{1}{2} = \frac{3}{5}$ **c.** $\frac{3}{4} = \frac{6}{8}$ **d.** $\frac{6}{9} = \frac{2}{3}$ **e.** $\frac{3}{4} > \frac{2}{3}$ **f.** $\frac{1}{3} > \frac{2}{5}$

g. $\frac{7}{8} > \frac{8}{10}$ **h.** $\frac{7}{14} > \frac{1}{2}$ **i.** $\frac{1}{3} < \frac{2}{6}$ **j.** $\frac{5}{8} < \frac{6}{7}$ **k.** $\frac{3}{2} < \frac{8}{10}$ **i.** $\frac{13}{15} < \frac{2}{3}$

4. Write the fractions in order, smallest to largest.

a. $\frac{1}{3}, \frac{1}{5}, \frac{1}{6}, \frac{1}{2}, \frac{1}{4}, \frac{1}{10}$ ______________________

b. $\frac{3}{4}, \frac{3}{5}, \frac{5}{8}$ ______________________

c. $\frac{2}{3}, \frac{4}{7}, \frac{7}{9}$ ______________________

⚠ Problems

Brain-teaser Jen has $\frac{1}{3}$ of a pizza. Tim has $\frac{2}{5}$. Which is the larger amount? ______

Brain-buster In a football stadium, $\frac{3}{7}$ of the crowd support the blue team and $\frac{1}{3}$ support the red team. The rest don't mind who wins; they are neutral. Arrange the crowd – red, blue or neutral – in order, by fraction, starting with the smallest.

______ < ______ < ______

Tricky fractions

Recap

Fractions show proportions of a **whole**.
Equivalent fractions represent the **same** proportion.

$\frac{3}{12} = \frac{1}{4}$

3 out of 12 of the dots are red.
This is the same as one quarter.

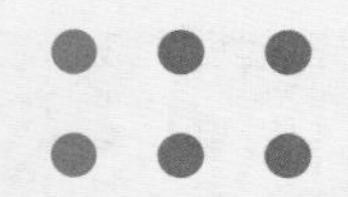

2 out of 6 of the dots are red.
This is the same as one third.

To compare and order fractions they must have a common denominator.

Which fraction is bigger, $\frac{1}{3}$ or $\frac{1}{4}$?

$\frac{1 \times 4}{3 \times 4} = \frac{4}{12}$ $\frac{1 \times 3}{4 \times 3} = \frac{3}{12}$

$\frac{1}{3} > \frac{1}{4}$

Revise

There were four cakes, but someone ate half of one of them.

There are now three and a half cakes.
We write this as $3\frac{1}{2}$.
There is a whole number and a fraction. This is called a **mixed number.**

Improper fractions have a numerator that is bigger than the denominator.
Look: $\frac{7}{2}$ is an improper fraction.

Look at the cakes and think about how many halves there are.
Each whole cake has two halves, so there are seven halves altogether. $\frac{7}{2} = 3\frac{1}{2}$

$\frac{7}{2}$ is the same as saying seven divided by two ... which is three and a half!

Tips

Do you know your numerators from your denominators?

- Remember that a fraction is a numerator divided by a denominator: $\frac{1}{2} = 1 \div 2$
- Converting improper fractions to mixed numbers is easy. $\frac{14}{3} = 14 \div 3 = 4$ r2 and as a mixed number $= 4\frac{2}{3}$
- Practise converting improper fractions into mixed numbers by writing down some improper fractions and, with a friend, challenge each other to make them into mixed numbers.

Talk maths

Each of these mixed numbers is equivalent (equal) to one of the improper fractions. Work with an adult or a friend to discuss which ones are equivalent, explaining your answers.

Mixed numbers

$3\frac{1}{4}$ $2\frac{2}{3}$ $2\frac{1}{2}$ $4\frac{1}{3}$ $3\frac{3}{5}$ $5\frac{1}{2}$ $1\frac{1}{5}$ $2\frac{3}{4}$ $2\frac{3}{10}$ $1\frac{8}{10}$ $4\frac{1}{10}$

Improper fractions

$\frac{6}{5}$ $\frac{41}{10}$ $\frac{13}{4}$ $\frac{18}{5}$ $\frac{11}{4}$ $\frac{11}{2}$ $\frac{13}{3}$ $\frac{18}{10}$ $\frac{5}{2}$ $\frac{8}{3}$ $\frac{23}{10}$

✓ Check

1. Change these mixed numbers to improper fractions.

a. $3\frac{1}{2}$ = ______ **b.** $2\frac{1}{4}$ = ______ **c.** $4\frac{1}{5}$ = ______ **d.** $1\frac{1}{3}$ = ______

e. $2\frac{2}{3}$ = ______ **f.** $2\frac{3}{4}$ = ______ **g.** $3\frac{4}{5}$ = ______ **h.** $8\frac{1}{2}$ = ______

2. Change these improper fractions to mixed numbers.

a. $\frac{3}{2}$ = ______ **b.** $\frac{4}{3}$ = ______ **c.** $\frac{5}{4}$ = ______ **d.** $\frac{6}{5}$ = ______

e. $\frac{11}{3}$ = ______ **f.** $\frac{7}{4}$ = ______ **g.** $\frac{15}{2}$ = ______ **h.** $\frac{13}{5}$ = ______

3. Insert =, < or > signs between each pair of fractions.

a. $\frac{3}{2}$ ______ $2\frac{1}{2}$ **b.** $\frac{4}{3}$ ______ $1\frac{1}{3}$ **c.** $\frac{7}{4}$ ______ $1\frac{1}{4}$ **d.** $\frac{13}{2}$ ______ $7\frac{1}{2}$

e. $6\frac{1}{4}$ ______ $\frac{25}{4}$ **f.** $3\frac{1}{2}$ ______ $\frac{8}{2}$ **g.** $\frac{10}{3}$ ______ $2\frac{2}{3}$ **h.** $3\frac{1}{5}$ ______ $\frac{12}{5}$

⚠ Problems

Brain-teaser A pizza shop sells pizzas as whole pizzas and in portions of half a pizza, a third of a pizza or a quarter of a pizza.

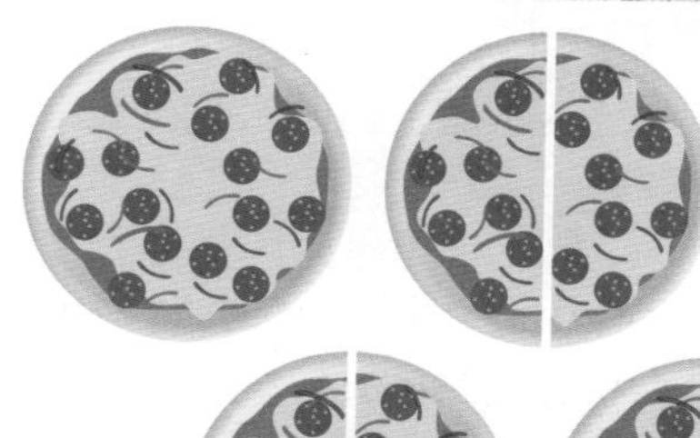

Ali orders five pizza halves, and Joanne orders seven pizza thirds.

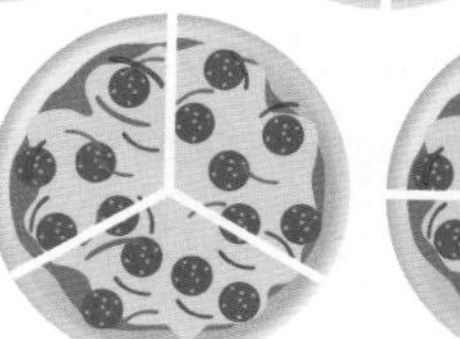

Who has ordered the most pizza? ____________________

Brain-buster The pizzeria chef has made four whole pizzas. Robin orders $\frac{13}{4}$ pizzas.

How much pizza will be left over? ____________________

Adding and subtracting fractions

Recap

We can easily compare fractions by giving them the same denominator.

$\frac{1 \times 4}{3 \times 4} = \frac{4}{12}$ $\frac{1 \times 3}{4 \times 3} = \frac{3}{12}$ so $\frac{1}{3} > \frac{1}{4}$

We can simplify fractions by dividing the top and bottom by a common factor.

$\frac{8 \div 2}{10 \div 2} = \frac{4}{5}$ $\frac{15 \div 5}{10 \div 5} = \frac{3}{2} = 1\frac{1}{2}$

Revise

To add and subtract fractions, they must have the same denominator.

Watch carefully…
To add $\frac{1}{4}$ and $\frac{3}{8}$, first find the lowest common denominator. This is 8.
Next, convert each fraction to give it a denominator of 8.

$\frac{1 \times 2}{4 \times 2} = \frac{2}{8}$ $\frac{3}{8}$ is ok

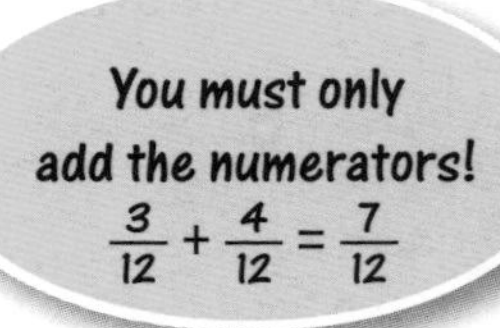

You must only add the numerators!
3/12 + 4/12 = 7/12

Then, add the new fractions:

$\frac{2}{8} + \frac{3}{8} = \frac{5}{8}$

Taking away is exactly the same – you only subtract the numerators.

$\frac{7}{10} - \frac{3}{10} = \frac{4}{10}$

Tips

- When you add fractions your answer might be an improper fraction. No problem!
  For example, $\frac{2}{3} + \frac{5}{6}$
  Lowest common denominator = 6
  $\frac{2}{3} = \frac{4}{6}$ $\frac{5}{6}$ is ok $\frac{4}{6} + \frac{5}{6} = \frac{9}{6}$
  As a mixed number that is $1\frac{3}{6}$. This can be simplified to $1\frac{1}{2}$.

Say simple additions and subtractions of fractions with common denominators, and challenge a friend to say if you are right or wrong.

$\frac{6}{7} - \frac{5}{7} = \frac{1}{7}$

Right!

$\frac{1}{5} + \frac{2}{5} = \frac{4}{5}$

Wrong! It is $\frac{3}{5}$.

✓ Check

1. Add these fractions.

a. $\frac{1}{5} + \frac{2}{5} =$ ______ **b.** $\frac{2}{7} + \frac{3}{7} =$ ______ **c.** $\frac{1}{4} + \frac{3}{4} =$ ______ **d.** $\frac{3}{5} + \frac{2}{5} + \frac{1}{5} =$ ______

2. Subtract these fractions.

a. $\frac{2}{6} - \frac{1}{6} =$ ______ **b.** $\frac{5}{8} - \frac{2}{8} =$ ______ **c.** $\frac{3}{4} - \frac{1}{4} =$ ______ **d.** $\frac{13}{20} - \frac{7}{20} =$ ______

3. Convert and add these fractions.

a. $\frac{1}{2} + \frac{1}{4} =$ ______________________

b. $\frac{1}{2} + \frac{1}{3} =$ ______________________

c. $\frac{1}{3} + \frac{1}{6} =$ ______________________

d. $\frac{4}{5} + \frac{1}{6} =$ ______________________

4. Convert and subtract these fractions.

a. $\frac{1}{2} - \frac{1}{4} =$ ______________________

b. $\frac{1}{2} - \frac{1}{3} =$ ______________________

c. $\frac{1}{3} - \frac{1}{6} =$ ______________________

d. $\frac{4}{5} - \frac{1}{6} =$ ______________________

⚠ Problems

Brain-teaser There are two identical pizzas at a party. Jim eats $\frac{1}{6}$ of one and $\frac{1}{3}$ of the other.

How much pizza does Jim eat altogether? ________________

Brain-buster Emma bakes a cake. She eats $\frac{1}{3}$ of it and her brother eats $\frac{1}{4}$ of it.

How much is left? ________________

Multiplying fractions and whole numbers

Recap

When multiplying by a fraction we use the word **of**.

Find $\frac{1}{2}$ of 12, $\frac{1}{4}$ of 12, $\frac{2}{3}$ of 12, $\frac{1}{6}$ of 12.
Use the dots to help you.

Don't forget, multiplication works in any order. $\frac{1}{2} \times 6$ is the same as $6 \times \frac{1}{2}$.

$\frac{1}{2}$ of 12 = 6.
One quarter of 12 is 3.
$\frac{2}{3}$ of 12 is 8.
One sixth of 12 is 2.

Revise

$\mathbf{3 \times \frac{1}{3} = 1}$, because $\frac{1}{3} + \frac{1}{3} + \frac{1}{3} = \frac{3}{3}$

$\mathbf{5 \times \frac{1}{3}} = \frac{1}{3} + \frac{1}{3} + \frac{1}{3} + \frac{1}{3} + \frac{1}{3} = \frac{5}{3}$ (or $1\frac{2}{3}$)

So, we multiply the numerator by the whole number. Look:

$4 \times \frac{2}{3} = \frac{8}{3}$ $4 \times \frac{3}{5} = \frac{12}{5}$ $7 \times \frac{1}{2} = \frac{7}{2}$

Can you change these into mixed numbers?

If you have to multiply a mixed number by a whole number, multiply each part separately then add them together:

$\mathbf{3 \times 4\frac{1}{2}}$

$\mathbf{3 \times 4 = 12}$ and $\mathbf{3 \times \frac{1}{2} = 1\frac{1}{2}}$

So, $\mathbf{3 \times 4\frac{1}{2} = 12 + 1\frac{1}{2} = 13\frac{1}{2}}$

When the numerator and the denominator are the same you always have one whole!

Tips

Let's make sure we have mixed numbers, not mixed-up numbers!

- Keep your work neat and clear, and make your answers as simple as possible. Like this:
  What is $\mathbf{8 \times 2\frac{1}{5}}$?
  $\mathbf{8 \times 2 = 16}$ and $\mathbf{8 \times \frac{1}{5} = \frac{8}{5}}$
  $\mathbf{8 \times 2\frac{1}{5} = 16\frac{8}{5}}$ (but remember, $\mathbf{\frac{8}{5} = 1\frac{3}{5}}$)
  So, $\mathbf{8 \times 2\frac{1}{5} = 17\frac{3}{5}}$

Talk maths

Practise saying these calculations both ways:

$4 \times \frac{1}{2} = \frac{4}{2} = 2$

$\frac{1}{2} \times 4 = 2$

Four times one half equals four halves, which equals two.
Half of four equals two.

Now try saying these:

$9 \times \frac{1}{3} = \frac{9}{3} = 3$ $\frac{1}{3} \times 9 = 3$

Try this with some other numbers and fractions.

✓ Check

1. **Complete these multiplications.**

   a. $\frac{1}{2}$ of 10 = ______ b. $\frac{1}{4}$ of 8 = ______ c. $\frac{2}{3} \times 9 =$ ______ d. $\frac{1}{2} \times 20 =$ ______

2. **Write the answers as mixed numbers.**

   a. $10 \times \frac{1}{4} =$ ______ b. $3 \times \frac{1}{2} =$ ______ c. $4 \times \frac{3}{7} =$ ______ d. $20 \times \frac{1}{6} =$ ______

3. **Find the answers.**

   a. $2 \times 3\frac{1}{2} =$ ______ b. $3 \times 1\frac{1}{2} =$ ______ c. $2\frac{1}{6} \times 5 =$ ______ d. $20 \times 1\frac{1}{3} =$ ______

⚠ Problems

Brain-teaser A farmer chops up trees to make logs for fires. If each tree makes $6\frac{1}{2}$ logs, how many logs will she get from three trees?

____

Brain-buster It takes Ben $12\frac{1}{2}$ seconds to jog once around the school hall. If he keeps up a steady speed, how many laps will he complete in 100 seconds?

____

Converting simple decimals and fractions

Recap

A proper fraction is a proportion of one whole.

$\frac{1}{4}$ $\frac{1}{3}$ $\frac{1}{2}$ $\frac{2}{3}$ $\frac{3}{4}$ are all proper fractions.

Amounts less than 1 can also be represented by decimals.

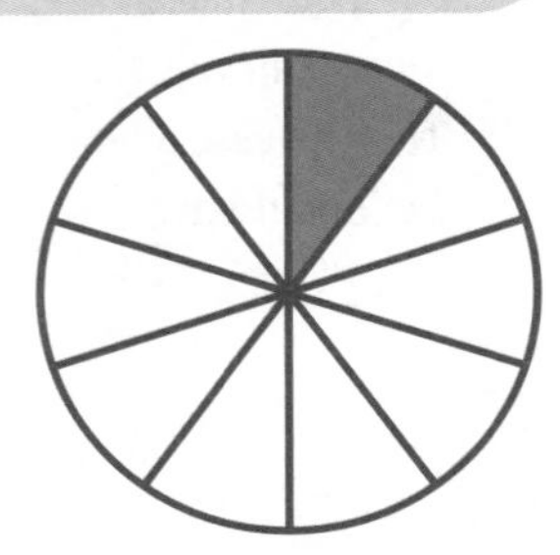

0.1 is one tenth = $\frac{1}{10}$

0.2 is two tenths = $\frac{2}{10}$

0.3 is three tenths = $\frac{3}{10}$

Revise

Any fraction can be written as a decimal. These are common ones:

Fraction	$\frac{1}{2}$	$\frac{1}{4}$	$\frac{3}{4}$	$\frac{1}{5}$	$\frac{1}{10}$
Decimal	0.5	0.25	0.75	0.2	0.1

There are 100 dots here.

50 of them are circled in blue.
As a fraction it is $\frac{50}{100}$ or $\frac{1}{2}$.
As a decimal it is 0.5.

Any decimal can be written as a fraction. For example:

25 of the dots are circled in green.
As a fraction it is $\frac{25}{100}$ or $\frac{1}{4}$.
As a decimal this is 0.25.

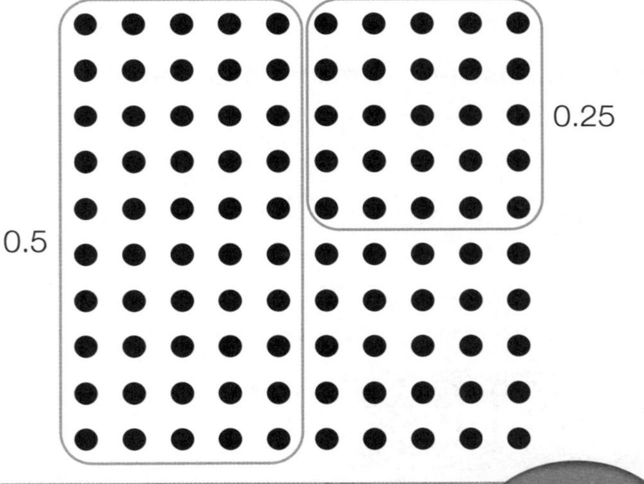

Tips

Time for some decimal tips!

- Remember that for a decimal the first column is tenths, the second column is hundredths, and the third column is thousandths.
- We read decimals aloud using numbers zero to 10.
  We say 0.5 is **zero point five**.
  We say 0.75 is **zero point seven five**.

100s	10s	1s	0.1s	0.01s	0.001s

Talk maths

Remember to read the decimals aloud. Try other fractions too! What will $\frac{4}{5}$ be? What will $\frac{6}{10}$ be?

Use this chart to make you familiar with the decimal equivalents for these common fractions. Work with a friend and test each other.

Fraction	$\frac{1}{2}$	$\frac{1}{4}$	$\frac{3}{4}$	$\frac{1}{5}$	$\frac{1}{10}$	$\frac{1}{3}$	$\frac{2}{3}$
Decimal	0.5	0.25	0.75	0.2	0.1	0.33	0.66

✓ Check

1. Complete these charts:

a.

Fraction	$\frac{1}{10}$	$\frac{2}{10}$	$\frac{3}{10}$	$\frac{4}{10}$	$\frac{5}{10}$	$\frac{6}{10}$	$\frac{7}{10}$	$\frac{8}{10}$	$\frac{9}{10}$	$\frac{10}{10}$
Decimal	0.1	0.2								

b.

Fraction	$\frac{1}{5}$	$\frac{2}{5}$	$\frac{3}{5}$	$\frac{4}{5}$	$\frac{5}{5}$
Decimal	0.2	0.4			

2. Change these fractions to decimals.

a. $\frac{1}{2}$ = ________ **b.** $\frac{3}{4}$ = ________ **c.** $\frac{1}{10}$ = ________

3. Change these decimals to fractions.

a. 0.25 = ________ **b.** 0.7 = ________ **c.** 0.4 = ________

⚠ Problems

Brain-teaser Adjith wins a competition. As his prize he can take $\frac{3}{4}$ of a bowl of sweets, or 0.7 of the sweets. Which will give him more sweets? ________

Brain-buster A pizza is cut into four equal slices. Gemma eats one slice and says that she has eaten 0.2 of the pizza. Is she right? Explain your answer.

____

____

Decimal fractions

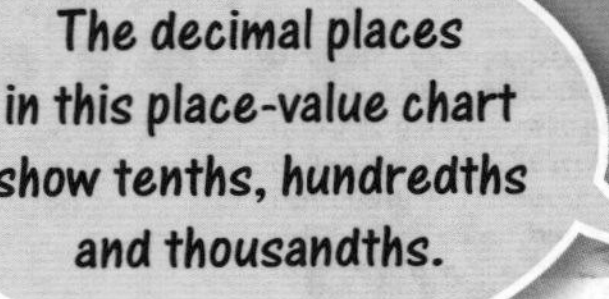

The decimal places
in this place-value chart
show tenths, hundredths
and thousandths.

Recap

Our number system uses **powers of 10**.
We sometimes call this 100s, 10s and 1s.

346 is three hundred and forty-six.
Between zero and 1 we use decimal fractions.
0.125 is **zero point one two five.**

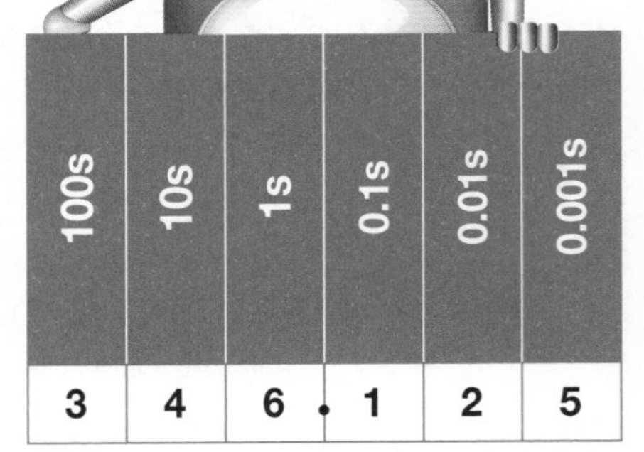

Revise

Can you see how many places the digit 1 moves each time?

A decimal fraction is a way of writing a fraction that has a power of 10, such as 10, 100, 1000, as its denominator.

When we divide a number by 10, 100 or 1000, we move the numbers to the right.

Fraction name	Fraction	Decimal fraction	Decimal name
one tenth	$\frac{1}{10}$	0.1	zero point one
one hundredth	$\frac{1}{100}$	0.01	zero point zero one
one thousandth	$\frac{1}{1000}$	0.001	zero point zero zero one

Here are some decimal fractions:

$\frac{7}{10} = 0.7$ $\frac{31}{100} = 0.31$ $\frac{418}{1000} = 0.418$

Tips

- When dividing by a power of 10, move the digits one place to the right for each power of 10.

Denominator of fraction	10	100	1000
The digit 1	0.1	0.01	0.001

Practise using the correct words for decimal fractions and decimal places.

$\frac{45}{100}$ *is forty-five hundredths* as a fraction.

For 0.45 as a decimal fraction we say *zero point four five.*

Say these numbers as fractions and also as decimal fractions.

0.2 **0.32** **0.523** **0.204** **0.8** **0.641** **0.09**

Complete this chart.

Fraction name	Fraction	Decimal fraction	Decimal name
five tenths			
twenty-three hundredths			
four hundred and thirty-five thousandths			
			zero point three
			zero point eight six
			zero point five zero seven
	$\frac{8}{10}$		
	$\frac{132}{1000}$		
		0.39	
		0.104	

Problems

Brain-teaser In a group of 100 children, 87 are right-handed. Write this number as a fraction

______________________ and as a decimal ______________________

Brain-buster In a group of 1000 children, 235 do not like cheese. What proportion of the children do like cheese?

As a fraction ______________________ As a decimal ______________________

Numbers with three decimal places

Recap

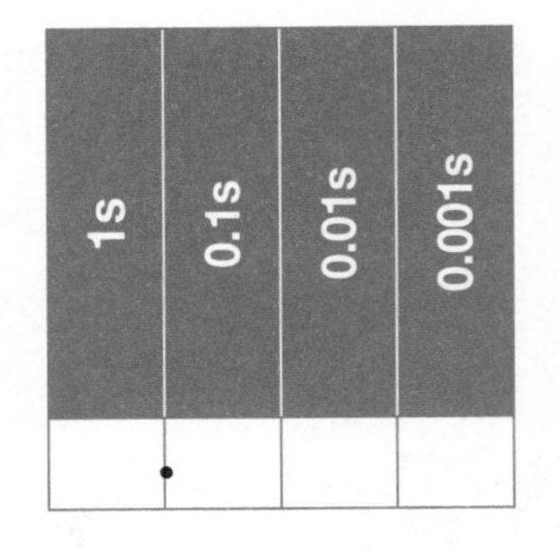

1s	0.1s	0.01s	0.001s
•			

Decimal places show the decimal fraction for numbers between zero and 1.

A decimal fraction is a way of writing a fraction that has a power of 10, such as 10, 100, 1000, as its denominator.

$$\frac{123}{1000} = \mathbf{0.123}$$

We say **one hundred and twenty-three thousandths as zero point one two three**.

Say each statement aloud.

Revise

We read decimals using digit names: 0.428 is **zero point four two eight**.

Tenths are bigger than hundredths.
Hundredths are bigger than thousandths. Look:

$0.6 > 0.5$ $0.431 > 0.429$ $0.1 > 0.099$

$0.28 < 0.3$ $0.739 < 0.81$ $0.4 < 0.515$

$$\frac{1}{10} > \frac{1}{100} > \frac{1}{1000}$$

Let's try to clear up different types of decimals.

- Think about what tenths, hundredths and thousands are:
  There are ten tenths in a whole.
  There are one hundred hundredths in a whole, but ten hundredths in one tenth.
  There are one thousand thousandths in a whole, but ten thousandths in one hundredth.

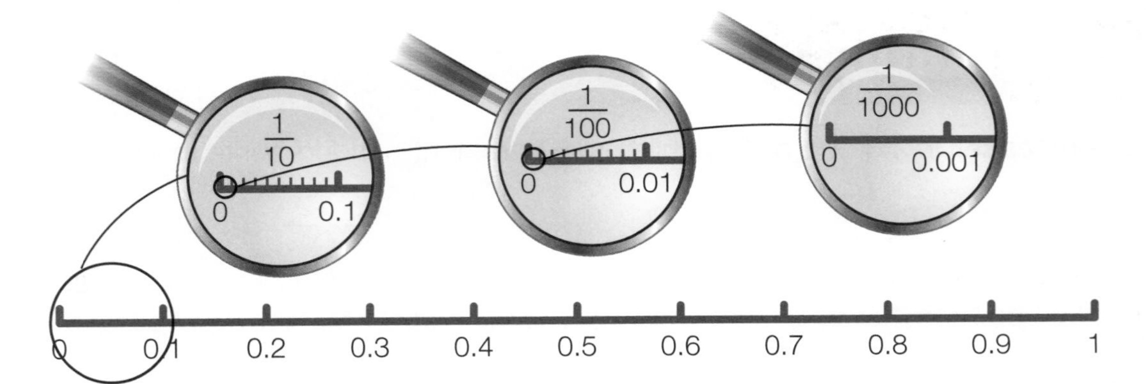

Talk maths

Prepare a short presentation to explain decimal fractions to a friend or an adult you know. Explain the differences between tenths, hundredths and thousandths, and how you can write these using words, decimals or fractions.

✓ Check

1. Write these in decimals using numerals.

a. zero point four six five ________

b. zero point two zero four ________

2. Write these in decimal fractions using numerals.

a. six tenths ________

b. twelve hundredths ________

c. three hundred and twenty-five thousandths ________

3. Write these decimals in words.

a. 0.395 ________________________________

b. 0.602 ________________________________

c. 0.005 ________________________________

4. a. Arrange these decimals in order, smallest to largest.

0.5 0.75 0.146 0.807 0.084 0.999 0.002 0.327

____

b. Now position them on the number line.

0 ———————————————————————————— 1

⚠ Problems

Brain-teaser Scientists can use microscopes to measure very tiny things. Circle the bugs that are less than three hundredths of a centimetre long:

Bug A: 0.101cm Bug B: 0.029cm Bug C: 0.009cm Bug D: 0.031cm

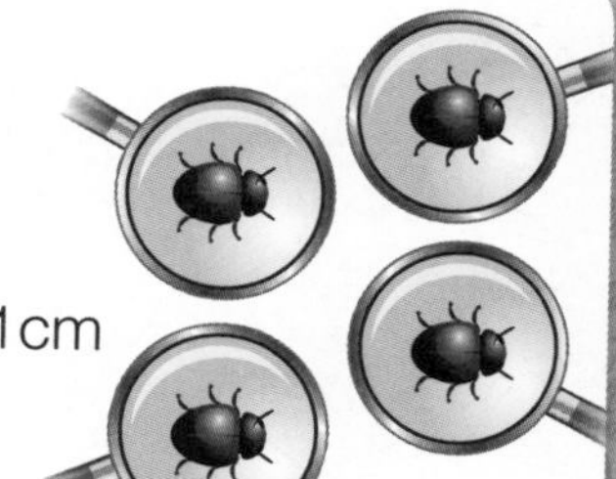

Brain-buster Arrange the bugs in order, smallest to largest.

____

Rounding decimals

Recap

0.235 has three decimal places: two tenths, three hundredths and five thousandths. 4.7 has one decimal place: seven tenths.

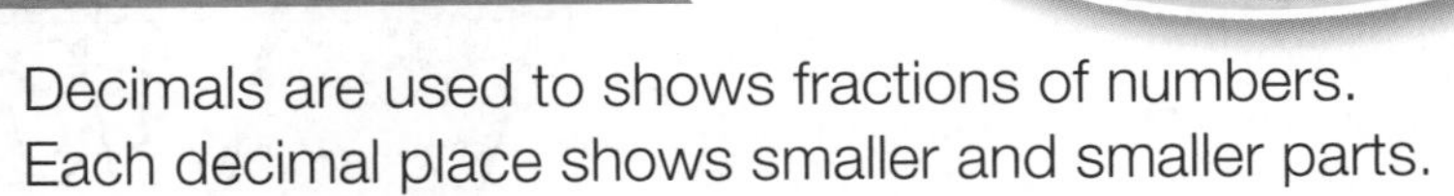

Decimals are used to shows fractions of numbers.
Each decimal place shows smaller and smaller parts.

Tenths **Hundredths** **Thousandths**

Revise

When rounding to the nearest tenth, you must look at the hundredths. 0.05 or more, round up. Less than 0.05, round down.

We can round decimals to the nearest whole number, it's easy:
0.5 or more, round up. Less than 0.5, round down.

0.65 rounds up to 1 | 2.34 rounds down to 2 | 0.723 rounds up to 1
8.058 rounds down to 8 | 4.629 rounds up to 5 | 0.255 rounds down to 0

Sometimes we want to be more accurate, and round numbers to one decimal place.

0.65 rounds up to 0.7 | 2.34 rounds down to 2.3 | 0.723 rounds down to 0.7
8.058 rounds up to 8.1 | 4.629 rounds down to 4.6 | 0.255 rounds up to 0.3

Tips

- Beware! The same number can be rounded off differently, depending on whether you round it to the nearest whole number, or to one decimal place.

7.49

To the nearest whole number: **7**

To the nearest tenth: **7.5**

7.51

To the nearest whole number: **8**

To the nearest tenth: **7.5**

Talk maths

Play *Don't Diss My Decimal*. Two or more people can play.
Take turns to think of any decimal with two or three decimal places and secretly write it down. Challenge others to ask questions about it to guess what it is. How fast can they discover your decimal?

0.475
Does it round up to the nearest whole number?
Does it round down to one decimal place?
Does it have thousandths?

✔ Check

1. Round these decimals to the nearest whole number.

a. 0.8 ______ **b.** 1.7 ______ **c.** 4.5 ______ **d.** 0.4 ______

e. 0.625 ______ **f.** 7.489 ______ **g.** 12.32 ______ **h.** 7.08 ______

2. Round these decimals to one decimal place.

a. 0.83 ______ **b.** 0.77 ______ **c.** 0.45 ______ **d.** 0.838 ______

e. 5.625 ______ **f.** 4.089 ______ **g.** 12.75 ______ **h.** 7.023 ______

3. Explain why rounding decimals is useful, but also why it might cause problems.

____

____

⚠ Problems

Brain-teaser A shopkeeper always rounds the money he makes each day to the nearest pound. Complete this chart.

Day	Monday	Tuesday	Wednesday	Thursday	Friday	Saturday
Money	£52.14	£45.61	£60.13	£46.50	£72.24	£35.51
Rounded						

Brain-buster For the week above, can you decide if the rounded amount is more or less than the money the shopkeeper actually has? Explain your answer.

____

Simple percentages

Recap

As a decimal it is 0.35: zero point three five.

$\frac{35}{100}$ is a fraction. We can say **35 over 100 or 35 out of 100**.

Fractions with a denominator of 100 are very important. They are also called percentages.

Revise

Per cent means parts of a hundred or out of 100.

Look at the 100 grid. 65 of the 100 squares are shaded – this is 65%.

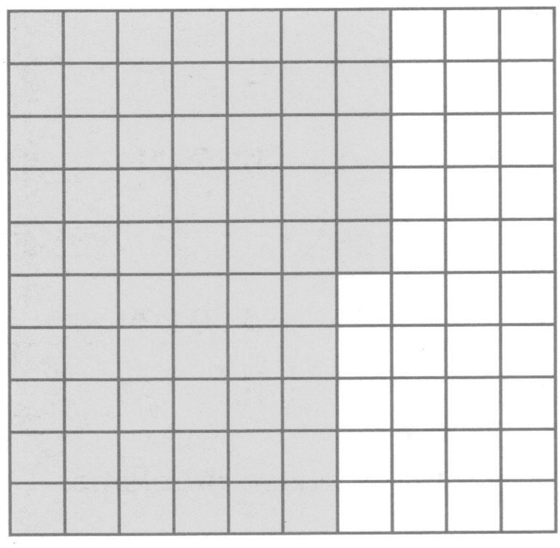

$0.65 = \frac{65}{100} = 65\%$

We use the words *per cent* and use the symbol % to represent it.

We can also simplify fractions as percentages:

50% $= \frac{50}{100} = \frac{1}{2}$ So, **50%** $= \frac{1}{2}$

Tips

Try to learn off by heart the percentage equivalents of easy fractions.

- Because fractions can also be turned into decimal fractions, they can also be percentages. Here are some you should know:

Fraction	$\frac{1}{2}$	$\frac{1}{4}$	$\frac{1}{10}$	$\frac{2}{10}$	$\frac{1}{5}$	$\frac{2}{5}$	$\frac{65}{100}$	$\frac{3}{4}$	$\frac{1}{1}$
Decimal	0.5	0.25	0.1	0.2	0.2	0.4	0.65	0.75	1.0
Percentage	50%	25%	10%	20%	20%	40%	65%	75%	100%

Talk maths

Look at the different fractions, decimals and percentages in the box.

$\frac{1}{2}$ 0.1 20% 0.3 1% $\frac{3}{4}$

67% $\frac{1}{5}$ 0.45 1.4 100% 0.8

For each one read it aloud, then say what it would also be as a decimal, percentage or fraction.

✔ Check

1. Link the correct percentage to its equivalent fraction and decimal.

10%	0.37	$\frac{1}{2}$
25%	1.0	$\frac{1}{10}$
30%	0.1	$\frac{1}{4}$
37%	0.3	$\frac{100}{100}$
50%	0.75	$\frac{3}{10}$
60%	0.25	$\frac{3}{5}$
75%	0.5	$\frac{37}{100}$
100%	0.6	$\frac{3}{4}$

⚠ Problems

Brain-teaser There are 30 children in a class. Half of the class have school dinners and $\frac{1}{10}$ have sandwiches. The rest of the class go home for lunch.

How many children have school dinners? ____________________

What percentage of the class have sandwiches? ____________________

Brain-buster In the above class, how many children go home for lunch? ____________________

What percentage of the class is this? ____________________

Length and distance

Recap

We usually measure longer distances in metres and kilometres, and shorter lengths in centimetres and millimetres.

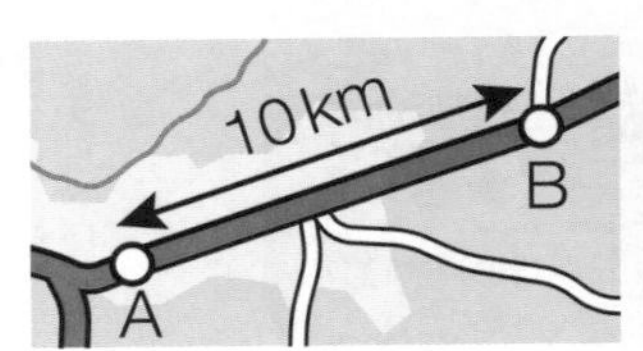

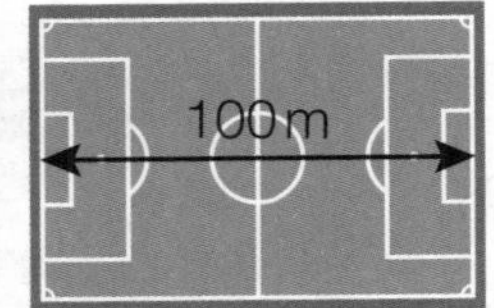

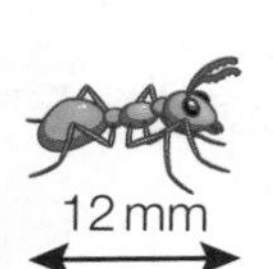

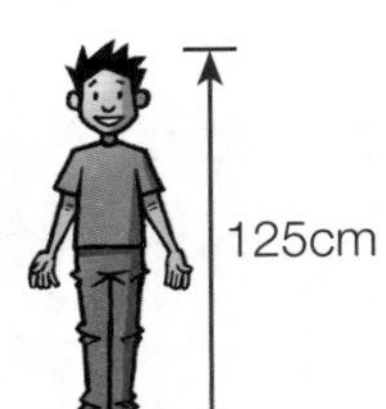

Abbreviations:
10mm = 1cm
100cm = 1m
1000m = 1km

Revise

Beware: when adding lengths together they must have the same units!

We sometimes use imperial units too:

0 cm 5 10 15
0 ins 1 2 3 4 5 6

A six-inch ruler is around 15cm.

1mm = 0.1cm
1cm = 0.01m
1m = 0.001km

TINY TOWN 5 MILES OR 8.05 KM

And a mile is around 1.61km.

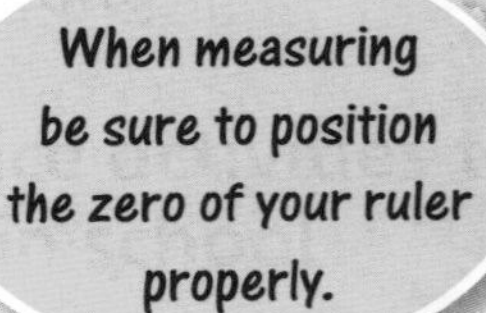

Tips

When measuring be sure to position the zero of your ruler properly.

- Use this guide to help you convert metric units.

Conversion	Operation	Example
mm to cm	÷ 10	12mm = 1.2cm
cm to m	÷ 100	256cm = 2.56m
m to km	÷ 1000	467m = 0.467km
cm to mm	× 10	3.5cm = 35mm
m to cm	× 100	1.85m = 185cm
km to m	× 1000	4.3km = 4300m

- For imperial units of length, just remember these two facts: **1 inch = 2.54cm** and **1 mile = 1.61km**

Talk maths

Can you measure to the nearest millimetre? Can you write all your measurements in millimetres?

You will need a ruler and a tape measure.

Working with an adult or a friend, choose a selection of different-sized objects. Write them in a list and then estimate their length. Write down your estimates and then swap lists. Measure the objects using the ruler and tape measure and compare answers.

✓ Check

1. Complete these conversion charts.

a.

mm	cm
10	
25	
52	
100	
	30
	17
	6
	0.2

b.

cm	m
100	
35	
450	
1000	
	80
	9
	0.9
	0.27

c.

m	km
1000	
250	
5350	
10,000	
	6
	4.5
	1.35
	0.004

2. Convert these imperial units to metric units. (Remember, 1 inch = 2.54cm and 1 mile = 1.61km.)

Imperial	Metric	Imperial	Metric	Imperial	Metric
1 inch		10 inches		100 miles	
5 inches		1 mile		300 miles	

⚠ Problems

Brain-teaser The desks in a classroom are all 120cm long. How long would a line of four desks be? Give your answer in metres. ____________________

Brain-buster An online map shows the length of sections of a cycle path. What is the total length of the path in km?

Path section	A–B	B–C	C–D	D–E	Total distance A–E
Distance	800m	750m	2.5km	1.3km	

Perimeter

Recap

Perimeter is the distance around the outside of a shape.

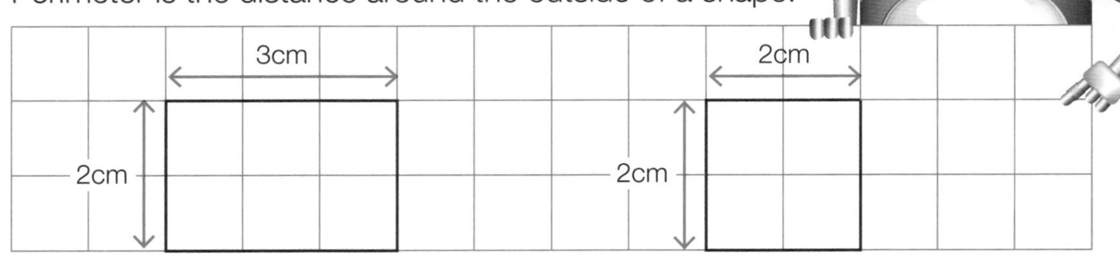

This rectangle has a perimeter of **3 + 3 + 2 + 2 = 10cm**.
We can also say **2 × 3 + 2 × 2 = 10cm**.

This square has a perimeter of **4 × 2 = 8cm**.

Revise

If all rectangles have a length and a width, then the perimeter can be calculated with a formula. In a formula letters are used instead of numbers.

P* = 2*l* + 2*w or ***P* = 2(*l* + *w*)**
Perimeter = 2 × **l**ength + 2 × **w**idth
The perimeter of this rectangle is *P* = 2 × 4 + 2 × 2 = 12cm.

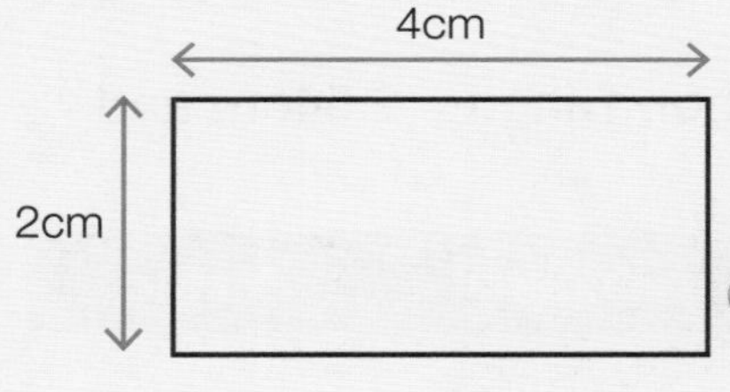

The formula for a square is easier, because all the **s**ides are the same length.
P* = 4*s
P = 4 × 3 = 12cm

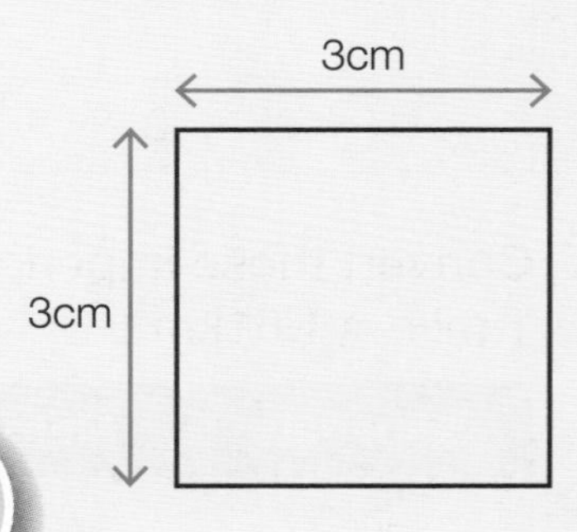

Tips

- **Composite** shapes are made of different shapes. Be very careful how you calculate the perimeter of this shape, which was made by joing a square and a rectangle. You might have to work out some measurements!

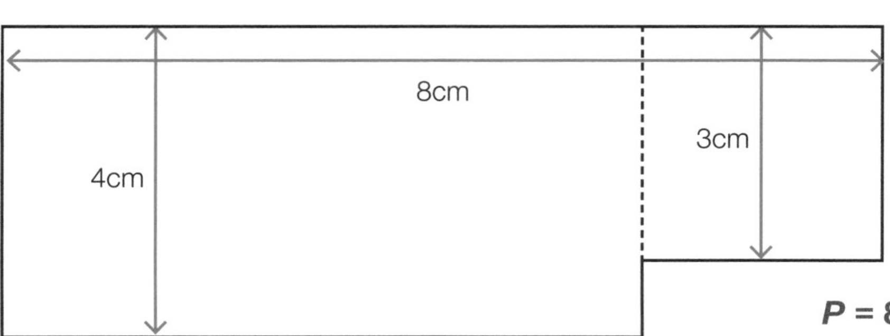

***P* = 8 + 3 + 3 + 1 + 5 + 4 = 24cm**

Talk maths

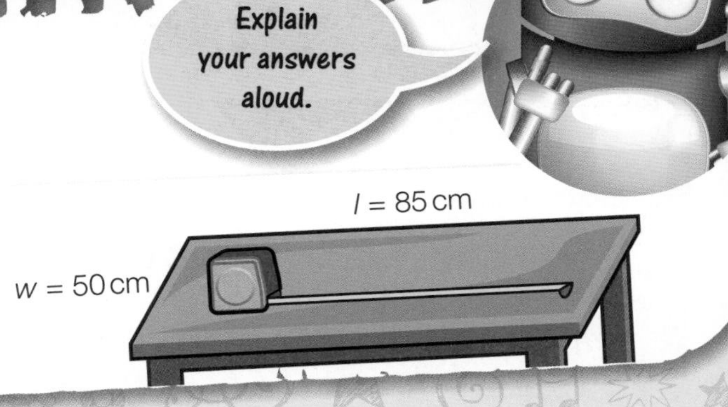

Practise estimating and measuring the perimeter of different rectangles and squares around your home. Try to calculate them to check your measuring.

✓ Check

1. Calculate the perimeter of these shapes.

a. 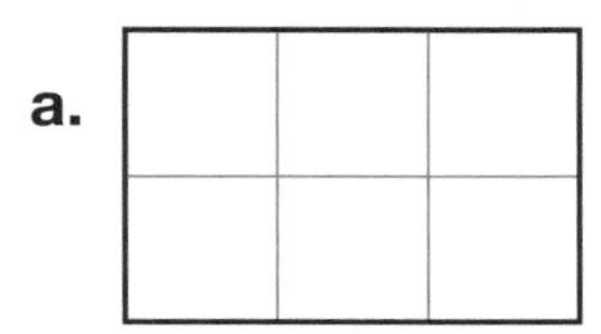 ______

b. 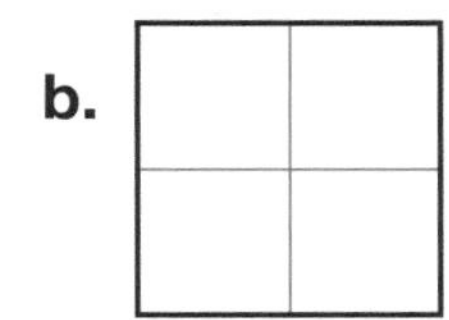 ______

c. ______

2. Calculate the perimeter of these composite shapes.

a. 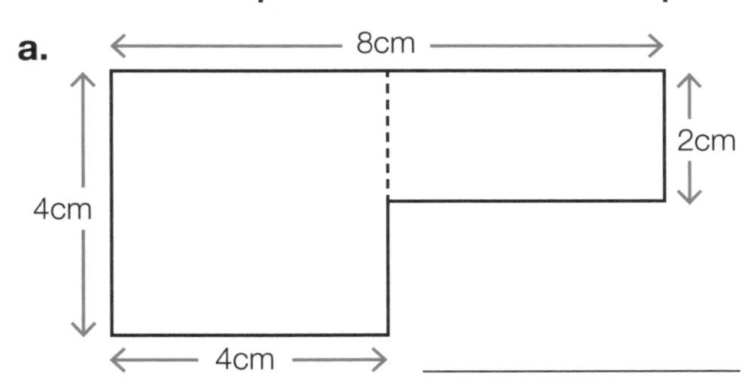

______

b. 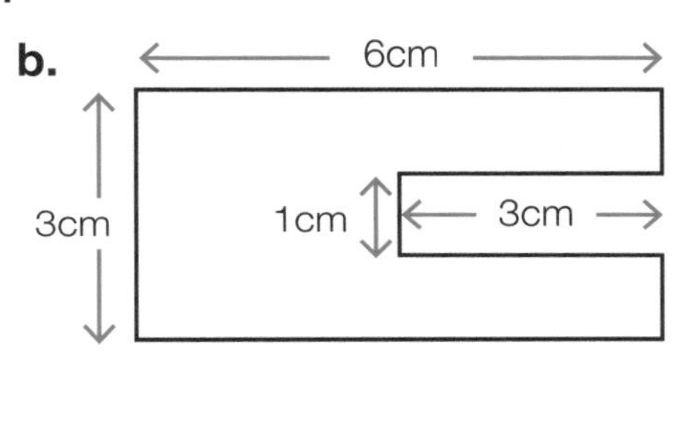

______

3. Complete this chart.

Shape	Formula	Length	Width	Perimeter
Rectangle	$P =$	6mm	3mm	
Square	$P =$	2.5mm	2.5mm	

4. Find the width of each of these shapes.

a. 6cm + 2w = 8cm, w = ______

b. 4m + 2w = 20m, w = ______

⚠ Problems

Brain-teaser Tina's garden is 7m long and has a perimeter of 20m.

How wide is the garden? ______

Brain-buster Tables in a classroom are 1m long and 0.5m wide.

What would be the perimeter of three tables pushed end to end? ______

Area

Recap

Area is measured in **square units**.

This square is 1cm long and 1cm wide.
Its area is $1cm^2$.

We can count squares to calculate simple areas.

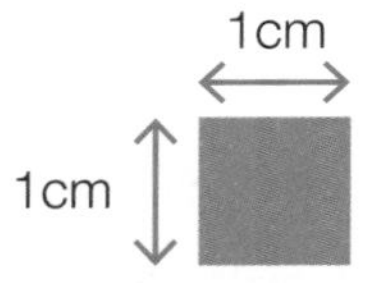

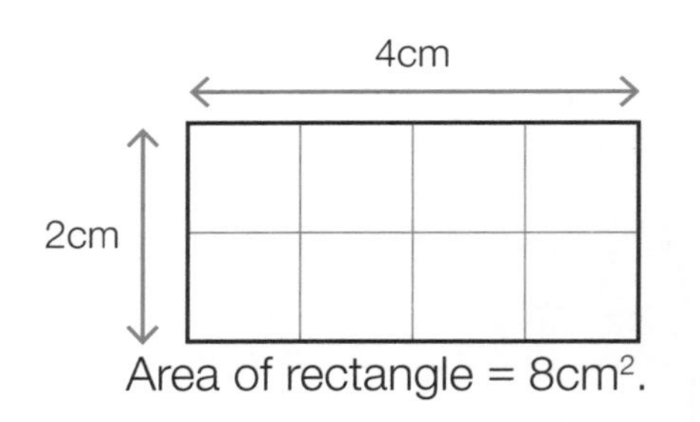

Area of rectangle = $8cm^2$.

Revise

The formula for calculating the **A**rea of rectangles and squares is the **l**ength times the **w**idth.
$A = l \times w$

The area of this rectangle is
$A = 5 \times 3 = 15cm^2$.

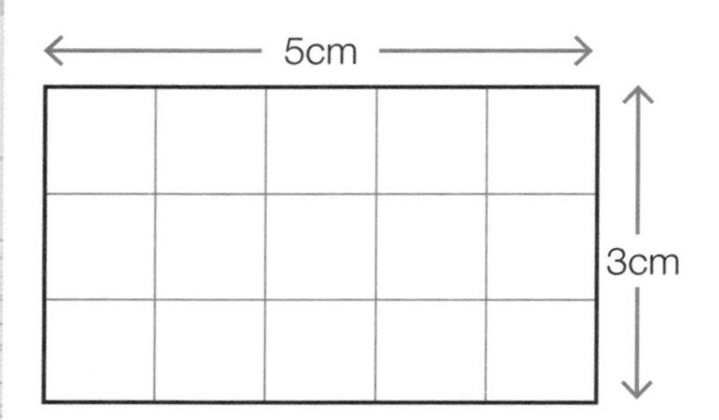

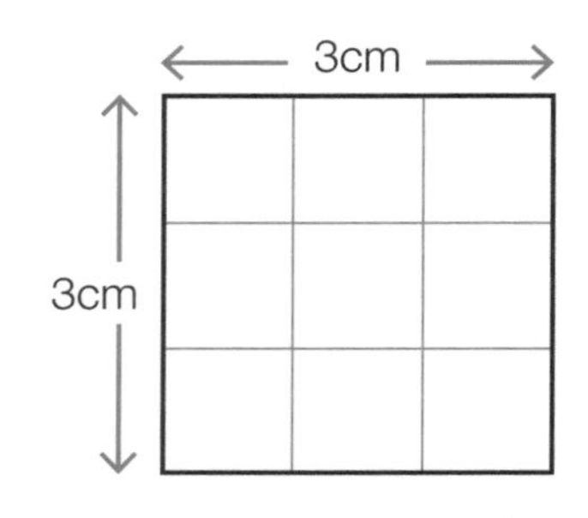

For squares, the length and the width are the same. The area of this square is $A = 3 \times 3 = 9cm^2$.

This field is 40m long and 30m wide.
The area of the field is
$A = 40 \times 30 = 1200m^2$.

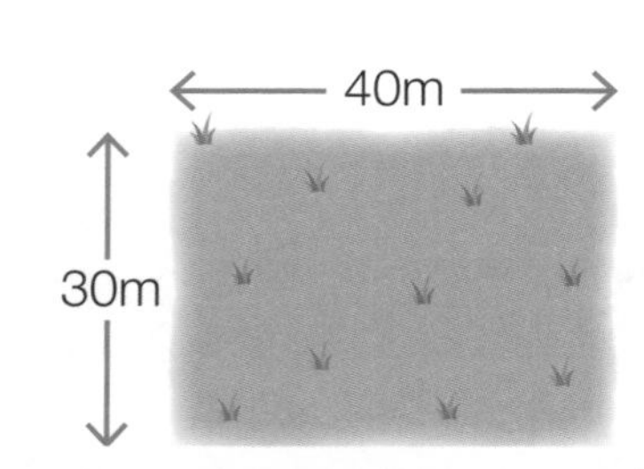

Tips

- If an irregular shape is drawn on $1cm^2$ paper we can still estimate its area. Just count the squares!

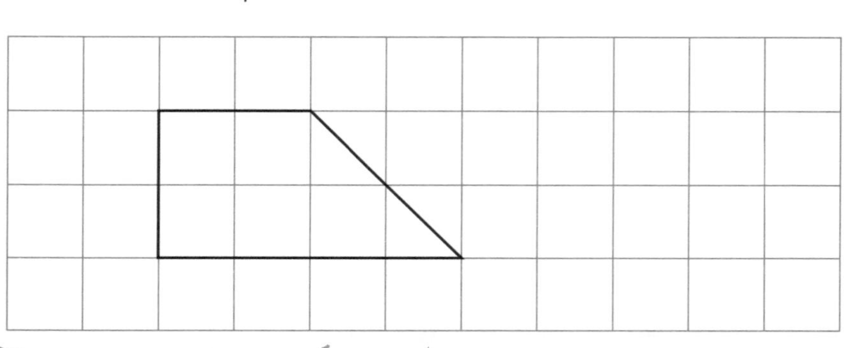

Talk maths

There should be more than one answer for each rectangle. How many can you find?

Look at these areas, and then tell an adult or a friend what the lengths and widths could be. Can they draw them?

Shape	Rectangle	Rectangle	Rectangle	Square	Square	Square
Area	$6cm^2$	$12cm^2$	$30cm^2$	$9cm^2$	$16cm^2$	$25cm^2$

✔ Check

1. Estimate the areas of these shapes assuming each square is $1cm^2$.

a. 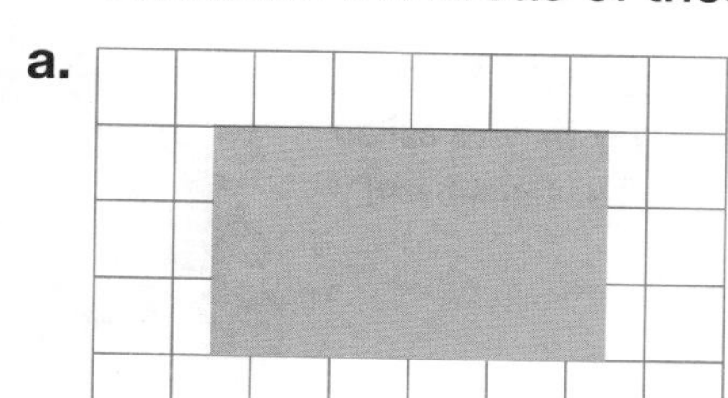

b. 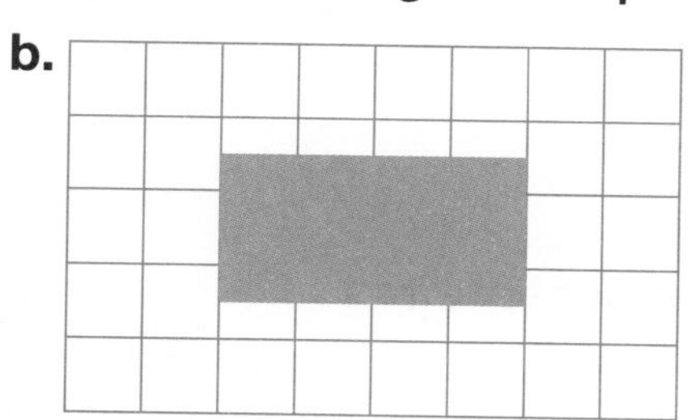

c. 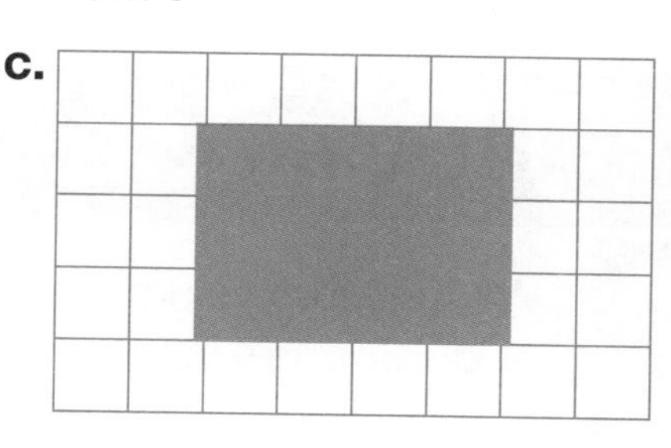

______________ ______________ ______________

2. Calculate the areas of these shapes.

a. A rectangle with length 12cm and width 9cm. ______________

b. A square with side length 7m. ______________

c. A rectangle with length 25m and width 12m. ______________

3. Circle which shape has the larger area.

a. A rectangle, length 5cm and width 1cm OR a square of side 2cm?

b. A rectangle, length 8cm and width 3cm OR a square of side 5cm?

c. A rectangle, length 7m and width 5m OR a square of side 6m?

d. A rectangle, length 17km and width 9km OR a square of side 12km?

⚠ Problems

Brain-teaser Annie's garden is 3m long and 5m wide. It was all grass but Annie has cut a square flowerbed, with side length of 1m, in the centre. What area of grass is left? ______________

Brain-buster A farmer plants nine potatoes in every square metre of earth.

How many potatoes will she grow in a field 120m long and 75m wide? ______________

Mass, capacity and volume

Recap

Mass is sometimes measured in pounds and ounces.

Capacity is sometimes measured in pints and fluid ounces.

Revise

Maybe that's why they're called cube numbers!

We measure volume in **cubic centimetres** or **cubic metres**.
The **V**olume of this cube is
$V = 1 \times 1 \times 1 = 1\text{cm}^3$.

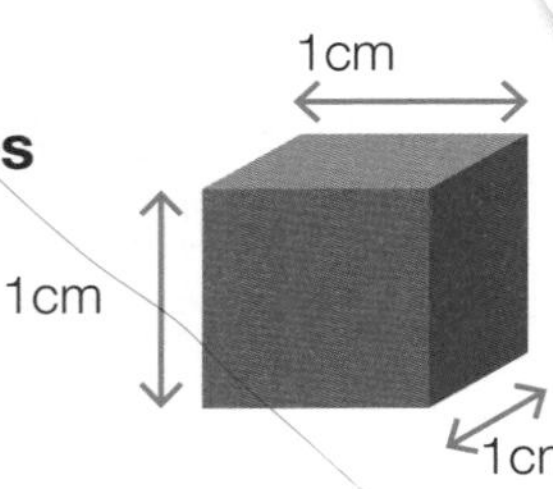

When we add masses together, the units must be the same.

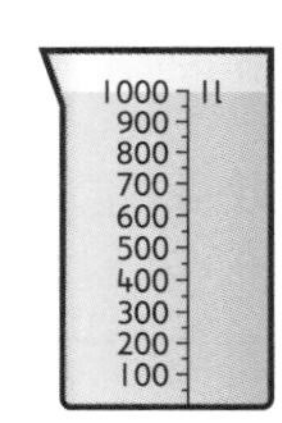

+

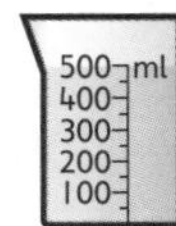

=

1500 ml
1000
500

1500ml

It's the same with capacity.

Tips

Let's make sure we know our units.

- Essential units for you to know:

Mass	1000g = 1kg
Capacity	1000ml = 1l

- To convert millilitres to litres, and to convert grams to kilograms, divide by 1000.
- To convert litres to millilitres, and kilograms to grams, multiply by 1000.

Talk maths

Volume is the amount of space an object takes up. Capacity is used to talk about how much something can hold.

Volume and capacity are have similarities.

Volume is usually measured in cubic units, such as cm³. One cubic centimetre equals one millilitre.

Collect some small objects from around your home, and work with an adult to try and estimate their volume. It would be great if you could use 1cm cubes to help you, but it is ok just to estimate.

✔ Check

1. Convert these masses.

Object	Grams	Kilograms
Child		50kg
Dog		12kg
Pencil	75g	
Book	408g	

2. Convert these capacities.

Object	Millilitres	Litres
Teapot		1.25l
Sink		8.5l
Mug	125ml	
Thimble	12ml	

3. Use the completed tables to answer these questions.

a. What is the mass of a pencil and a book? ______________________

b. What is the capacity of a mug and a thimble? ______________________

c. What is the mass of a child and a book? ______________________

d. What is the capacity of a teapot and a thimble? ______________________

⚠ Problems

Brain-teaser Hakan weighs 50kg and Aysha weighs 35kg. Their teacher weighs 65kg.
How much extra weight would their teacher need to add to make a seesaw balance if he sits on one end and the children on the other?

Brain-buster Françoise has a litre bottle of water. How much will she have left if she fills two thimbles and a mug? The capacity of one thimble is 12ml and of one mug is 125ml.

____

Time

Recap

Except for leap years – they have an extra day!

Seconds, **minutes**, **hours**, **days**, **weeks**, and **years** – these are all units of time. Months are a bit different because they are not all the same length.

60 seconds = 1 minute
60 minutes = 1 hour
24 hours = 1 day
7 days = 1 week
365 days = 1 year

Analogue clocks show 12-hour time.
Digital clocks can show **12-hour** or **24-hour time**.

Revise

Converting between different units of time is not difficult.

Minutes to seconds	× 60		Seconds to minutes	÷ 60
Hours to minutes	× 60		Minutes to hours	÷ 60
Days to hours	× 24		Hours to days	÷ 24
Weeks to days	× 7		Days to weeks	÷ 7
Years to days	× 365		Days to years	÷ 365

Remainders stay in the same units.
For example, change 400 days to years.
400 ÷ 365 = 1 remainder 35
So, **400 days = 1 year and 35 days**.

Or one year and five weeks. Can you see that?

Tips

- Be careful adding minutes that move on to a new hour. For example, if a school's lunch break starts at 12:30pm and lasts for 40 minutes, at what time will the break be over? There are 60 minutes in an hour, so 12:30pm + 40 = 1:10pm.
- If you are not sure, use a clock or watch to help you.
- Also, be aware of the 12-hour and 24-hour clocks. Remember that 1pm = 13:00, 6pm = 18:00, and so on.

Talk maths

Talk to as many people as you can about their birthday. Try to find out what time of day they were born.

Use a calendar to challenge your friends to work out how many days there are between their birthday and yours.

Look at timetables at bus stops or train stations. Can you find out how long different journeys take?

DID YOU KNOW?

10 years is known as a decade, 100 years is known as a century and 1000 years is known as a millennium.

✔ Check

1. Complete this chart to convert 12-hour clock times to 24-hour clock times.

12-hour	Midnight	2.10am	9.15am	Noon	3.30pm	9pm	11.59pm
24-hour							

2. Write these times in the unit shown.

a. 3 hours in minutes ________________

b. 2 days in hours ________________

c. 12 weeks in days ________________

d. 4 years in days ________________

e. 100 minutes in hours ________________

f. 100 hours in days ________________

g. 225 minutes in hours ________________

h. 500 days in years ________________

3. Add these times.

a. 8 hours and 25 minutes + 3 hours and 30 minutes ________________

b. 1 day 7 hours and 25 minutes + 3 days 5 hours and 16 minutes ________________

c. 2 days 17 hours and 45 minutes + 6 days 9 hours and 13 minutes ________________

⚠ Problems

Brain-teaser Rob's school bus journey takes 17 minutes. If he gets on the bus at 8:20am, what time will he arrive at school?

____

Brain-buster Gayle's new baby brother will be exactly one week old at 6pm today. Help Gayle to calculate how many seconds old her new brother will be at 6pm.

____

Money

Recap

Money shows us the cost of things.
We use pounds (£) and pence (p).

£1 = 100p

We show pence using decimals.

13 pounds and 65 pence = £13.65

If you use the £ sign you don't need to add a p at the end.

Revise

You can use all your number skills to solve money problems.

	Example	Try these:
Addition	£13.50 + £6.37 = £19.87	£7.32 + £2.50 =
Subtraction	£20.00 – £16.25 = £3.75	£10.00 – £8.85 =
Multiplication	£2.30 × 4 = £9.20	£5.25 × 5 =
Division	£9.00 ÷ 4 = £2.25	£4.50 ÷ 3 =
Fractions	$\frac{1}{2}$ of £16.50 = £8.25	$\frac{1}{4}$ of £20 =

Tips

- Calculations with money are just the same as using any decimals that have two decimal places.

£47.87 + £38.17

```
    4 7 . 8 7
  + 3 8 . 1 7
  -----------
    8 6 . 0 4
  -----------
    1 1   1
```

Answer: **£86.04**

Not sure about money? Maybe decimals can help.

Talk maths

Work with an adult or a friend. Use these items to create multi-step problems to challenge each other. Use written methods to calculate answers if you need to, but be sure to discuss your answers.

Example: If I buy two apples and a banana, how much change will I get from a £2 coin?

2 × 23 + 35 = 81p

2.00 − 0.81 = £1.19

✓ Check

1. Convert these pence to pounds.

Pence	200p	135p	6325p	9p	10,903p
Pounds					

2. Convert these pounds to pence.

Pounds	£4	£2.56	£0.12	£82	£403.20
Pence					

3. Complete these calculations (use written methods if necessary).

a. £4.52 + £3.25 = ________________

b. £12.35 + £9.80 = ________________

c. £10.00 − £8.30 = ________________

d. £45.45 − £3.72 = ________________

e. £3.35 × 2 = ________________

f. £14.08 × 5 = ________________

g. £50.00 ÷ 4 = ________________

h. $\frac{1}{2}$ of £15.40 = ________________

⚠ Problems

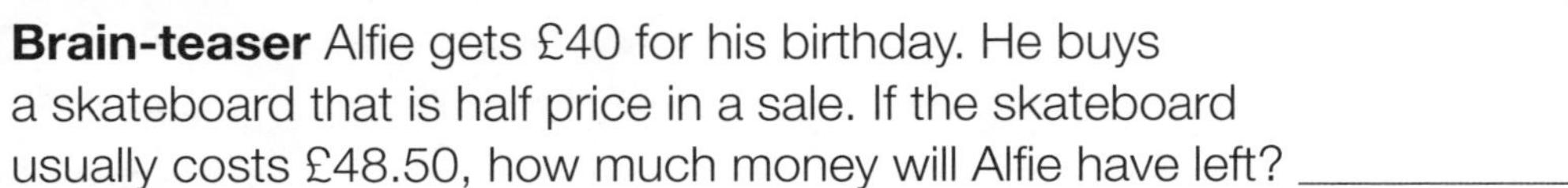

Brain-teaser Alfie gets £40 for his birthday. He buys a skateboard that is half price in a sale. If the skateboard usually costs £48.50, how much money will Alfie have left? ________________

Brain-buster A head teacher orders ten desks that cost £35.50 each and 20 chairs for £12.25 each. What will be the total cost? ________________

Angle facts

Recap

We measure angles with a protractor.

Revise

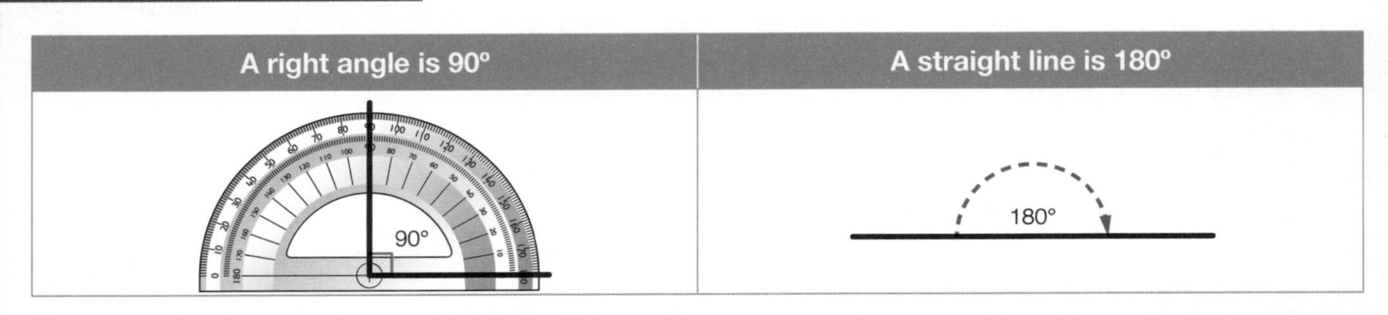

A right angle is 90°	A straight line is 180°
90°	180°

Acute angles are between 0° and 90°	Obtuse angles are between 90° and 180°
50°	160°

Angles greater than 180° are called ***reflex*** angles	A complete turn is 360°
200°	360°

Tips

- Make sure you can use a protractor properly. Check the size of each of the angles on this page.
- Think about it… two lines that make an acute angle on the inside will make a reflex angle on the outside. These two angles add up to 360°.

Stuck? Let's come at things from a different angle.

Talk maths

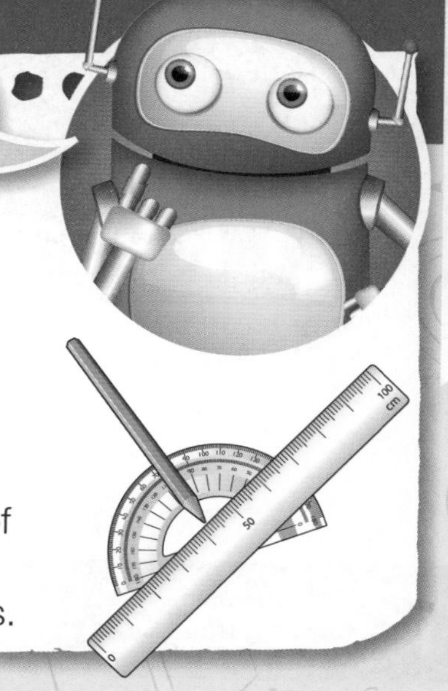
How accurate were your angle estimates?

You will need paper, a pencil, a ruler and a protractor.

First, explain to an adult how to use a protractor properly. Demonstrate the correct way to place it on an angle, and how to measure the angle using the correct readings.
Next, work together to become angle experts: draw a selection of angles and then estimate their size.
Finally, measure each one and compare them with your estimates.

✔ Check

1. Measure these angles. Write down their size and their name, such as *acute*.

a.

size ____________________

name ____________________

b.

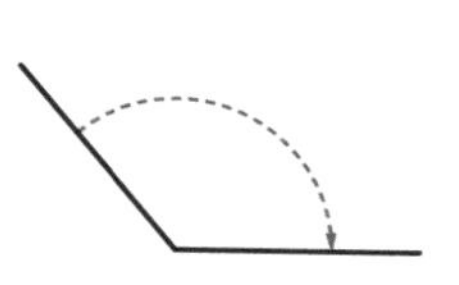

size ____________________

name ____________________

c.

size ____________________

name ____________________

2. Draw these angles, and then name them.

a. 90°

name ____________________

b. 300°

name ____________________

c. 15°

name ____________________

⚠ Problems

Brain-teaser Aaron only has a ruler and a pencil, but he can still say if an angle is acute, obtuse or reflex. How can he do this?

____

Brain-buster Amy draws a straight line and then draws another line to its middle to make two separate angles. She measures the acute angle to be 42°. She says that the obtuse angle must be 138°. Explain why she is right.

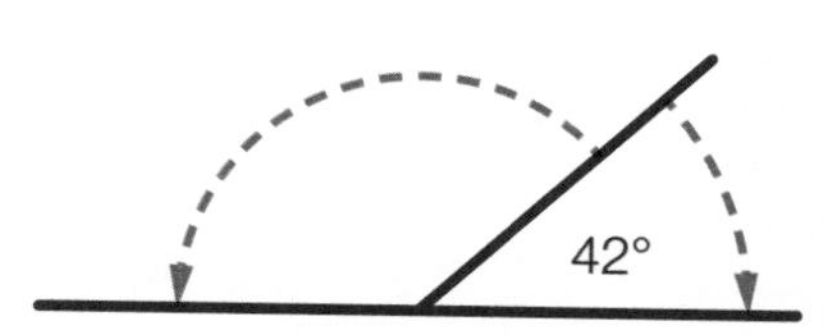
42°

____

Rotating angles

Recap

A right angle is 90°.
Two right angles make 180°.

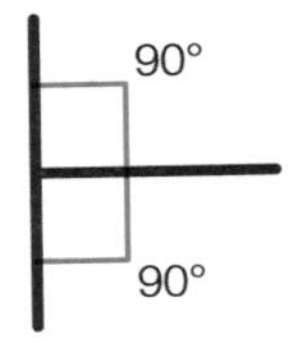

Revise

There are 360° in a complete rotation.

A complete rotation is four right angles.
90° + 90° + 90° + 90° = 360°

Each rotation of 90° is a right angle.

Rip the four corners off a piece of A5 paper and put them together. What do you notice?

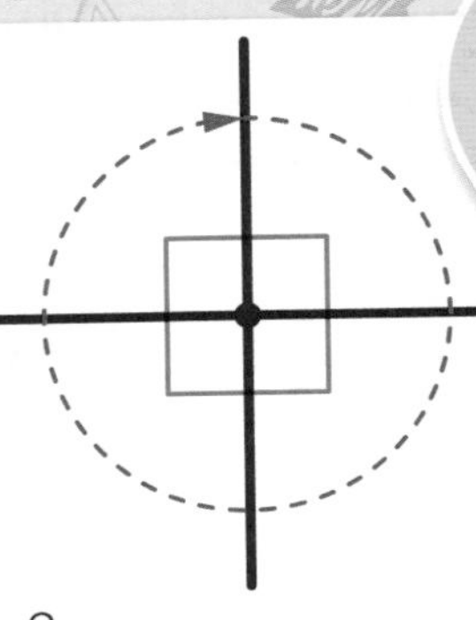

Imagine standing at the centre of this circle and turning ninety degrees four times.

Tips

- You can rotate clockwise or anticlockwise.

Use clocks to practise counting in right angles. This clock shows a clockwise rotation of 270°.

If you go anticlockwise the rotation is 90°.

270 + 90 = 360° (one complete turn)

Compasses have a 90° rotation between each point. North to east is 90° clockwise, or 270° anticlockwise.

North to south is 180° clockwise and 180° anticlockwise!

You can play a mini version of this activity using an action figure on paper!

Talk maths

You will need some chalk and a flat area of ground.

With a friend, find a safe and quiet place, and then use chalk to draw a large cross on the ground and mark the end of each line 1, 2, 3, 4.

Challenge each other to rotate the correct number of right angles. For example, *face number 2 and then rotate 90°*. Which number are you facing now?

✔ Check

1. Draw these angles using a protractor.

a. 90° **b.** 180° **c.** 270°

d. 360° **e.** 0°

⚠ Problems

Brain-teaser To get from midday to 9pm, how many right angles does the hour hand on a clock have to turn through? ____________

How many degrees is this? ____________

Brain-buster Michael stands facing south. He rotates 270° clockwise, and then he turns an amount. If he finishes facing west, how many degrees did he turn the second time, and in what direction? ____________

2D shapes

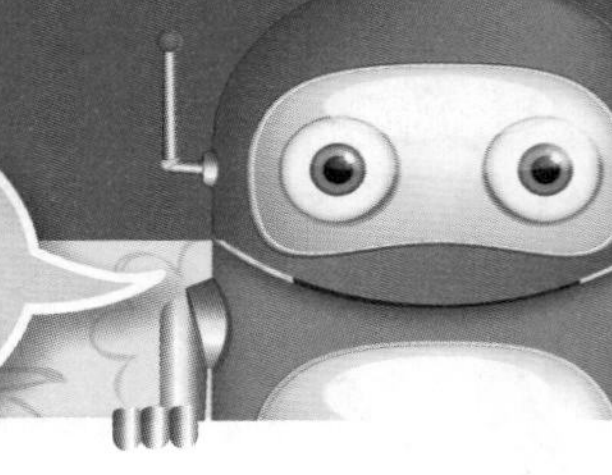

Each of the angles in a square is 90°. A circle has one side and no corners!

Recap

We say that different 2D shapes have different properties.

Triangle	Quadrilateral	Pentagon	Hexagon	Heptagon	Octagon
3 sides	4 sides	5 sides	6 sides	7 sides	8 sides

All of these shapes are **regular** – all the sides are the same length, and in each shape all of the angles are the same size.

Revise

Irregular shapes have the same number of sides and angles as their corresponding regular shapes, but the sides and angles are not identical.

Triangle	Quadrilateral	Pentagon	Hexagon	Heptagon	Octagon
3 sides	4 sides	5 sides	6 sides	7 sides	8 sides

Tips

- There are several types of quadrilateral that you also need to know about.

Square	Rectangle	Rhombus	Parallelogram	Kite	Trapezium
All sides equal, all angles 90°	Opposite sides equal, all angles 90°	All sides equal, opposite angles equal	Opposite sides equal, and parallel opposite angles equal	Adjacent sides equal	Only one pair of parallel sides

- **Adjacent** means next to.

Talk maths

Cover the names of the shapes on the opposite page. Try to identify each shape, telling a friend or an adult why you think it is that particular shape.
Then cover the shapes and try to describe the properties for each name.

DID YOU KNOW?

A nonagon is a 9-sided shape and a decagon is a 10-sided shape.

✓ Check

1. Name each shape, and say if it is regular or irregular.

a. 

b. 

c. 

d. 

2. Explain why a rectangle is not a regular quadrilateral.

3. Label these quadrilaterals.

a. ____________ b. ____________ c. ____________

d. ____________ e. ____________ f. ____________

⚠ Problems

Brain-teaser Write the length of the missing sides and the size of the missing angles on this rectangle.

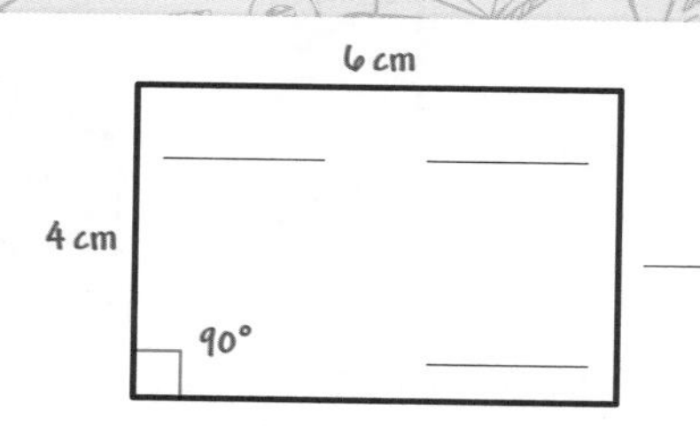

3D shapes

Recap

3D shapes have faces, edges and vertices.

A corner is a **vertex**.
The plural is **vertices**.

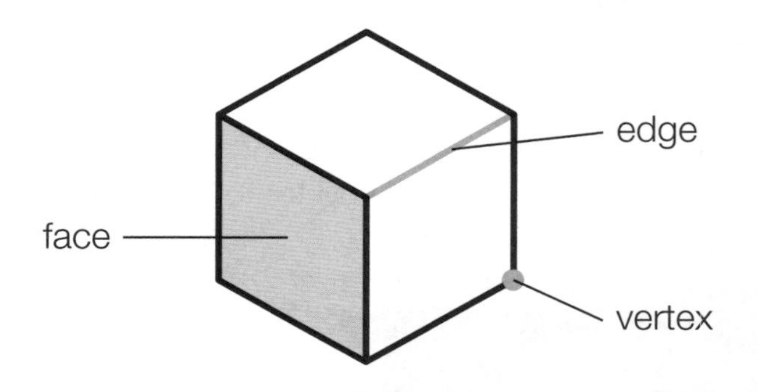

Revise

Irregular shapes have the same number of sides and angles as their corresponding regular shapes, but the sides and angles are not identical.

Shape						
Name	Cube	Cuboid	Cone	Sphere	Cylinder	Triangular prism
Faces	6	6	2	1	3	5
Edges	12	12	1	0	2	9
Vertices	8	8	0	0	0	6

Tips

- Drawing shapes to look 3D is called isometric drawing.
- The trick is to draw one end face, and then draw the edges as parallel lines.

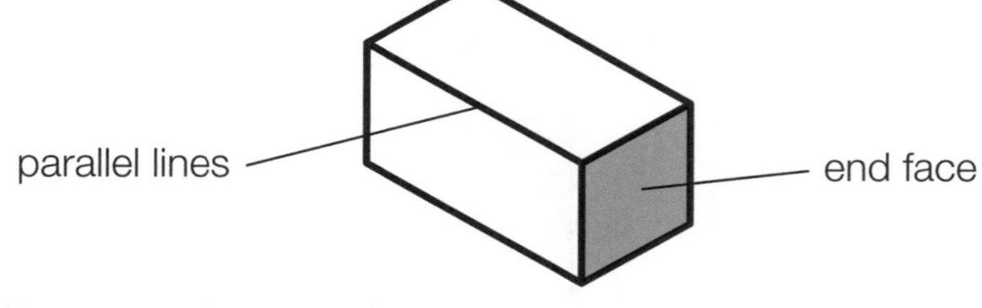

- Beware: spheres and cones are harder to draw!

You don't
need special
glasses to view
3D shapes!

Talk maths

With two or more people, play the *Yes/No* game for shapes.
Choose a person to be the drawer.
All other players will be the guessers.
The drawer draws a shape on paper, making sure the guessers don't see it.
The guessers then ask questions to find out which shape has been drawn.
The only answer the drawer can give is *yes* or *no*.

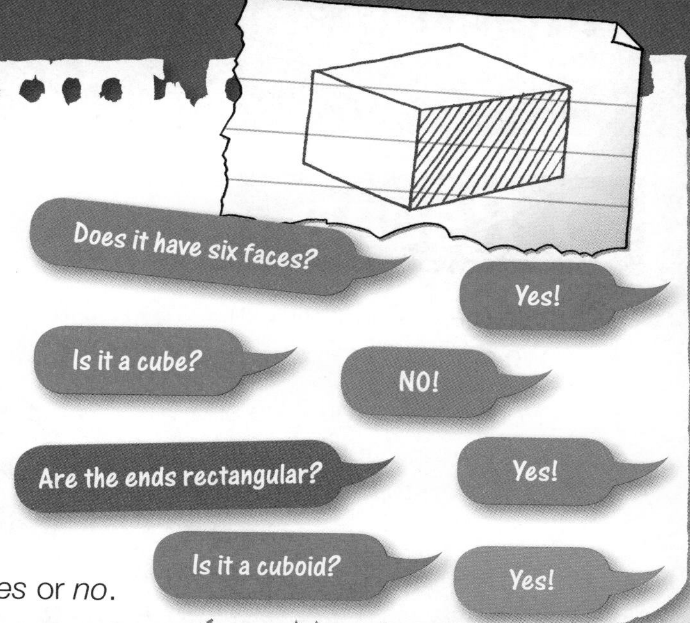

Does it have six faces?
Yes!
Is it a cube?
NO!
Are the ends rectangular?
Yes!
Is it a cuboid?
Yes!

✓ Check

1. Use a pencil and ruler to draw each of these shapes.

triangular prism	cylinder	cube	cuboid

2. Name the shapes from their descriptions below.

a. I have only one face. ____________________

b. I have six identical faces. ____________________

c. I have only one edge. ____________________

d. I have three faces. ____________________

e. I have five faces. ____________________

f. I have six faces, some different. ____________________

⚠ Problems

Brain-teaser Sanjay draws a 3D shape with a square end; each edge of the square is 4cm. The other sides of his shape are 6cm long. What shape has he drawn?

____

Brain-buster A cone and a cylinder are both 10cm long and both have a base with a diameter of 5cm. If they were hollow, which one could hold more water? Explain your answer.

____

____

Reflecting and translating shapes

Recap

We draw a grid with an x-axis and a y-axis, and we can plot the corners of polygons (2D shapes).

Points on the grid are shown with the x-coordinate first, and then the y-coordinate.

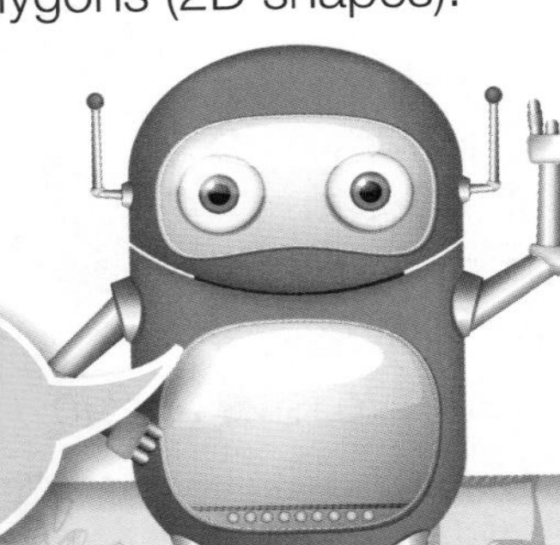

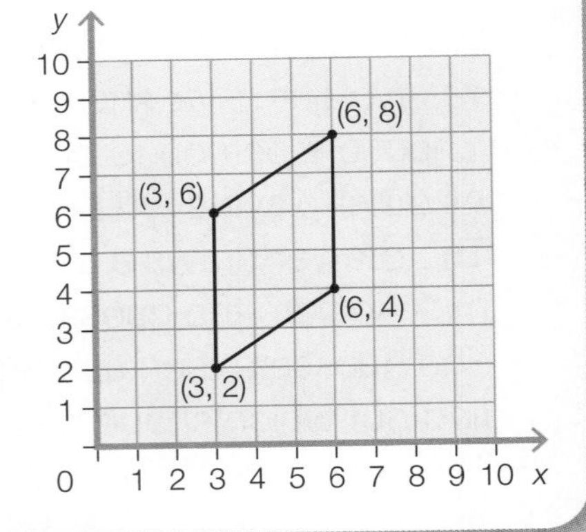

Revise

A has been reflected to A'.
The x-coordinate has not changed.

We can also reflect shapes.
Try reflecting the triangle.

We can also **translate** shapes.

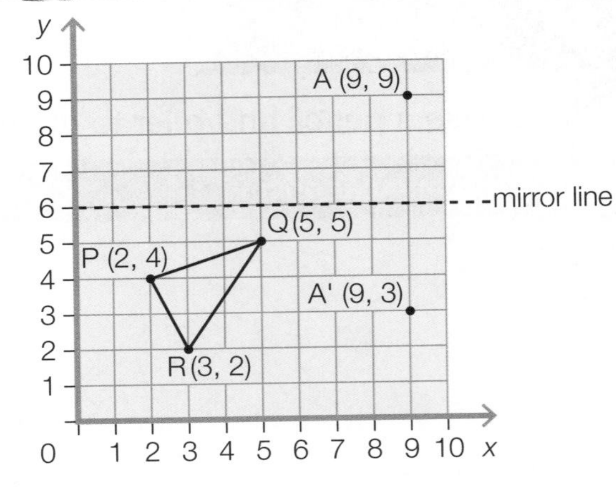

That is when all the points move the same distance.

The square has been translated (5, 4).

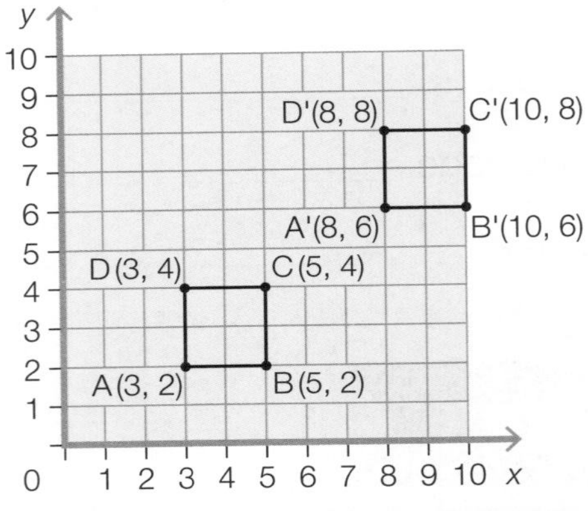

Tips

- **Reflections:**
  In a vertical mirror line, only the x-coordinates change.
  In a horizontal mirror line, only the y-coordinates change.
- **Translations:**
  All the x-coordinates should change by the same amount.
  And so should the y-coordinates.

Talk maths

Take turns to choose a point on the graph and say its coordinates. Challenge someone to reflect or translate it.

Reflect the point (2, 3) vertically.

Reflect the point (9, 1) horizontally.

Translate the point (0, 5) by (2, 3).

Translate the point (4, 4) by (-2, -3).

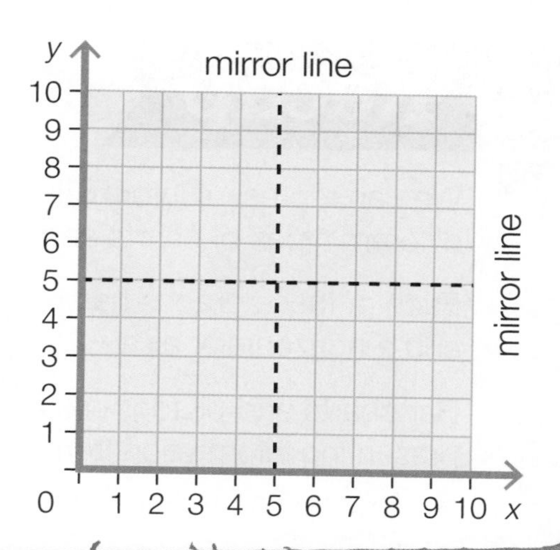

✔ Check

1. Look at the graph below. Using squared paper:

a. Reflect the square PQRS.

b. Write the coordinates of the new square.

c. Plot a triangle ABC: A (6, 5), B (9 ,6), C (7, 9).

d. Reflect the triangle.

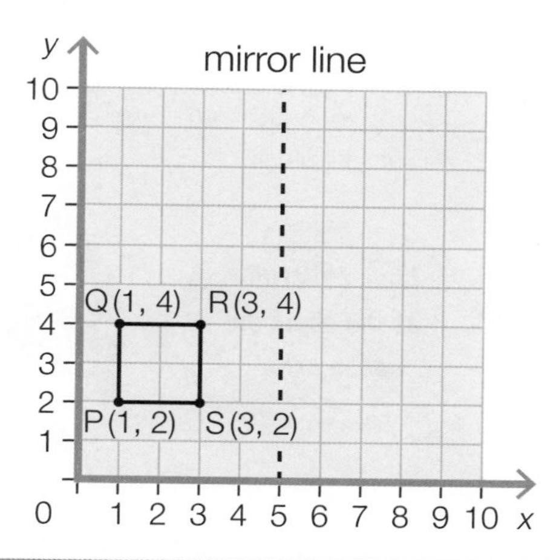

2. Look at the graph below. Using squared paper:

a. Translate the square WXYZ by (3, 2).

b. Write the coordinates of the new square.

c. Plot a triangle DEF: D (2, 7), E (4, 9), F (3, 5).

d. Translate it by (5, –4).

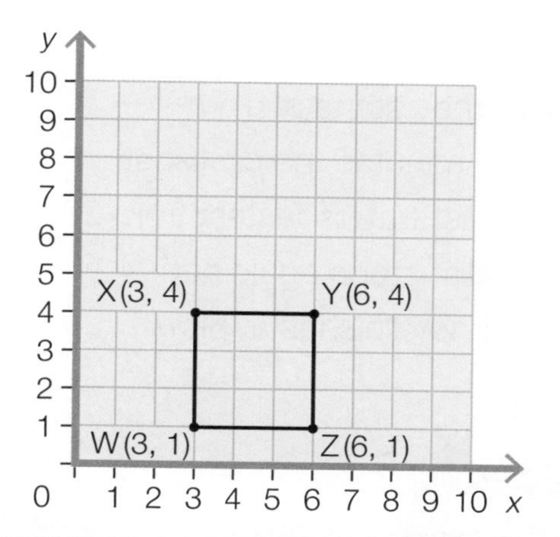

⚠ Problems

Brain-buster Gayle has worked out a method for calculating the coordinates of reflected shapes without drawing them. Explain her method.

____

____

Line graphs

Recap

We can represent information and data in different types of charts and graphs.

Each of these graphs has a vertical y-axis and a horizontal x-axis.

Bar charts and pictograms are useful for presenting information from surveys, such as:

- How do you travel to school?
- What is your favourite snack?
- Do you have any pets?

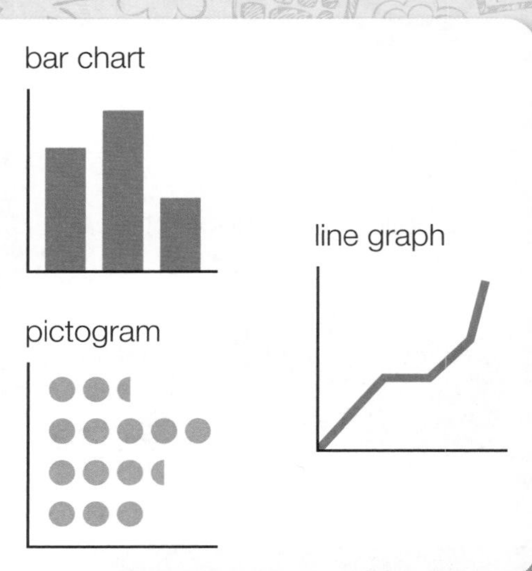

Revise

Line graphs are useful to show how things change over time, such as temperature, growth and speed.

This graph shows the time taken for an 8km cycle ride.

Find these bits of information on the graph:

- The journey starts at 1pm.
- After 20 minutes the cyclist stops for five minutes.
- The cyclist travels fastest from 25 minutes to 40 minutes.
- The cyclist stops again after 40 minutes.
- The journey finishes at 8km.

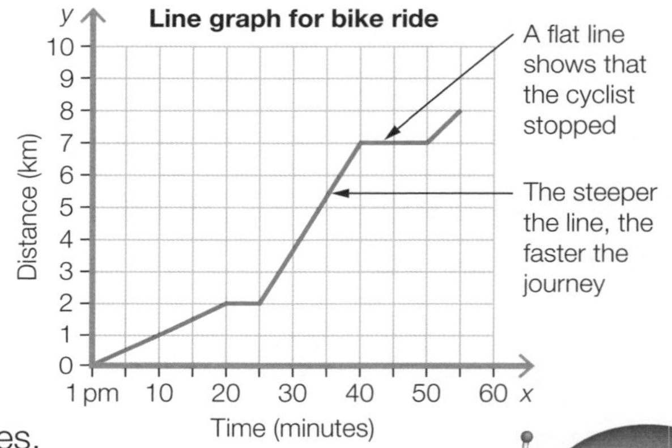

Line graph for bike ride

The steeper the line, the faster the journey
Look carefully at the scale on each axis.

Tips

- Line graphs can be used to estimate information.
- This graph shows the height a tree grew every two years for ten years.
- We can draw lines to show its height after three years.

When it comes to tips on graphs, I draw the line!

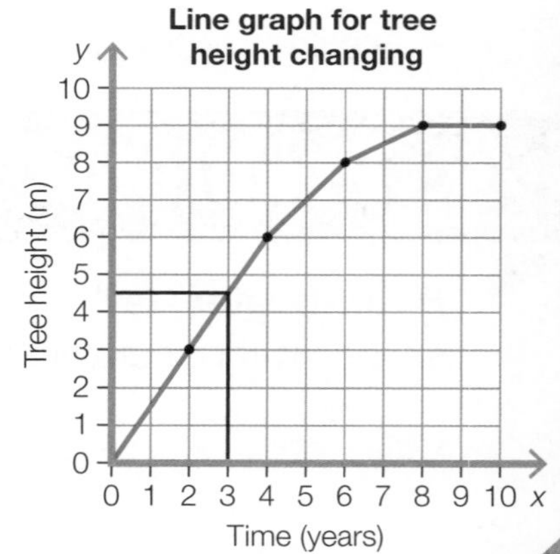

Line graph for tree height changing

Tree height (m)

Talk maths

This line graph shows the distance travelled by a lorry on a long journey.

Look at the graph with an adult and talk about what each part means.

What is each axis for?

What is the scale for each axis?

How long is the journey?

What is happening when the line is flat?

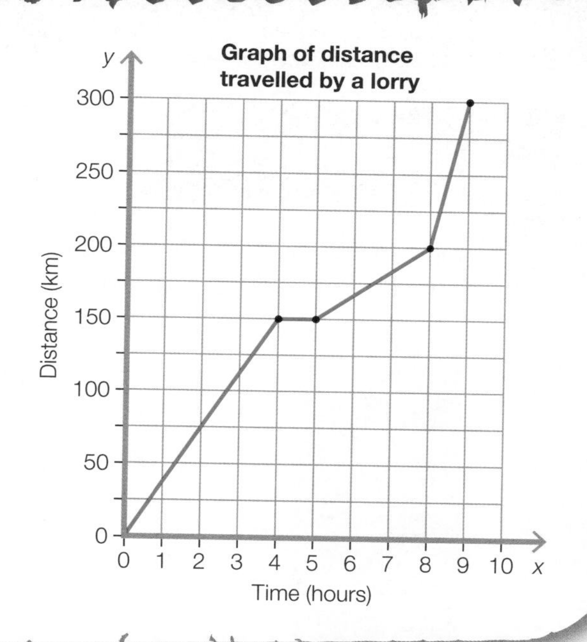

✔ Check

1. Philip grows a plant at home. He measures it at the end of each week and records its height in a chart.

Time (weeks)	1	2	3	4	5	6	7	8	9	10
Height (cm)	0	1	3	5	7	10	11	12	12	12

a. Draw a line graph to show how the plant has grown.

b. When was the plant 3cm high? ____________________

c. When did the plant stop growing? ____________________

d. Which week did the plant grow the most?

____

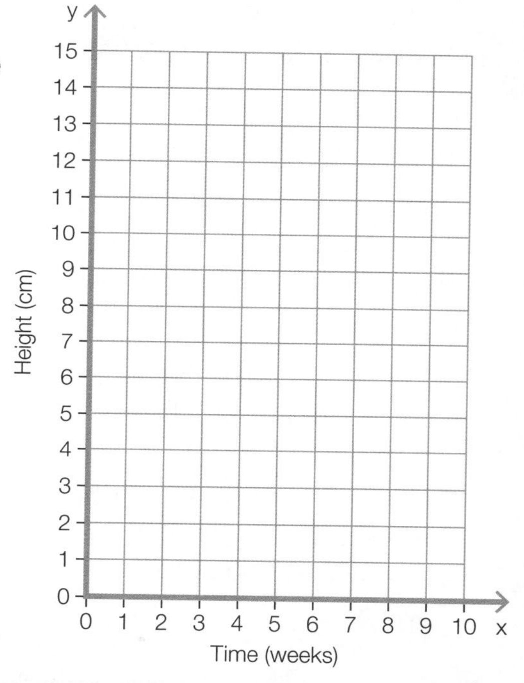

⚠ Problems

Brain-teaser What is the highest temperature?

____

When was temperature highest? ____________________

Brain-buster Find the difference between the highest and lowest temperatures.

____

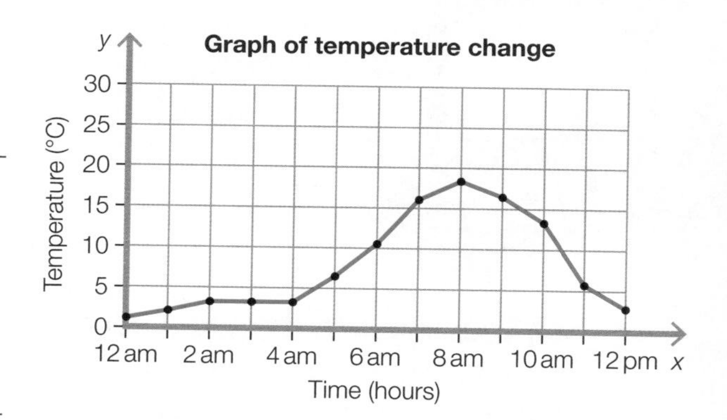

Tables and timetables

You can spot facts about planets and compare them too. Each column has different units.

Recap

Information is often presented in tables. This table provides complicated information about some planets in our solar system.

Planet	Diameter (km)	Day length (hours)	Orbit time (days)	Temp. (°C)
Mercury	4878	4223	88	167
Venus	12,104	2808	225	480
Earth	12,756	24	365	20
Mars	6794	24.5	687	–65

Revise

This timetable shows bus times from the bus station to the local school.

Check with an adult that you know how to read the timetable properly. These buses have a circular route. Can you see the halfway point?

Can you see the difference between the number 6 and 7 bus routes? Why wouldn't you go to the supermarket on the number 6 bus? What is the longest time it takes to get between two stops?

BUS TIMETABLE

Bus number	6	7	6	7	6	7
Bus station	10:00	10:15	10:30	10:45		
High Street	10:08	10:23	10:38	10:53		
Supermarket		10:30		11:00		
Train station	10:15		10:45			
Doctor's surgery	10:24	10:34	10:54	11:04		
School	10:27	10:37	10:57	11:07		
Doctor's surgery	10:30	10:40	11:00	11:10		
Train station	10:39		11:09			
Supermarket		10:45		11:15		
High street	10:46	10:52	11:16	11:22		
Bus station	10:54	11:00	11:24	11:30		

Can you spot the patterns for the bus times? Try filling in the last two columns.

Tips

Let your fingers do the walking!

- Use your fingers to help you trace along timetables and charts. Or if you have a ruler available, using this even better – it is easy to misread timetables and charts.

Talk maths

Ask an adult to provide you with a timetable. These are usually available for local buses and trains. You can get them from the station and they will probably be available online.

Use the timetables to plan a day trip.
Can you plan at least four journeys with as little time waiting as possible?

DID YOU KNOW?

There are over 5 billion bus journeys made in the UK each year.

✔ Check

1. Use the table on page 152 to answer these questions about our solar system.

a. Which is the largest planet? ______

b. Which is the coldest planet? ______

c. Which planet has the fastest orbit? ______

d. Which planets have a similar length of day? ______

______

2. **Use the bus timetable on page 152 to answer these questions.**

a. Which bus has a shorter journey from the bus station to the school? ______

b. How often does the number 6 leave the bus station? ______

c. How long is the journey from the doctor's surgery to the school? ______

d. Why do you think the number 6 round journey takes longer? ______

______

⚠ Problems

Look at the bus timetable on page 152.

Brain-teaser Trevor has a doctor's appointment at 10:45am.

Which bus should he catch from the bus station? ______

Brain-buster A number 7 bus arrives at school at 15:07.

What time did it leave the bus station? ______

English glossary

A

adjectives are sometimes called 'describing words' because they pick out features of nouns such as size or colour. They can be used before or after a noun, to give more detail. The red bus.

adverbs can describe the manner, time, place or cause of something. They tell you more information about the event or action.

adverbials are words or phrases that give us more information about an event or action. They tell you how, when, where or why something happened.

alliteration is the repetition of a consonant sound or letter in several words: beautiful black butterfly.

analogy is a comparison in which an idea is compared to something that is quite different. It compares the idea to something that is familiar to the reader. There are plenty more fish in the sea.

apostrophes:

- show the place of missing letters (**contraction**)
- show who or what something belongs to (**possession**).

assonance is the repetition of a vowel sound in several words: aggressive angry alligator.

B

brackets show parenthesis. They are placed around extra information in a sentence. Alex (who had got up late) ran all the way to school.

C

clauses are groups of words that must contain a subject and a verb. Clauses can sometimes be complete sentences.

- **main clause:** contains a subject and verb and makes sense on its own.
- **subordinate clause:** needs the rest of the sentence to make sense. A subordinate clause includes a conjunction to link it to the main clause.
- **relative clause:** is a type of subordinate clause that changes a noun. It uses relative pronouns such as who, which or that to refer back to that noun.

command tells someone to do something and ends with an exclamation mark or a full stop.

commas have different uses including:

- to separate items in a list
- to separate a fronted adverbial from the rest of the sentence
- to clarify meaning
- to show parenthesis.

common noun names something in general.

conjunctions link two words, phrases or clauses together. There are two main types of conjunction:

- **co-ordinating conjunctions** (and, but) link two equal clauses together.
- **subordinating conjunctions** (when, because) link a subordinate clause to a main clause.

consonants are most of the letters of the alphabet except the vowel letters a, e, i, o, u

contraction a shortened word where an apostrophe shows the place of missing letters.

co-ordinating conjunctions (and, but) link two equal clauses together.

D

dashes in pairs show parenthesis.

determiners go before a noun (or noun phrase) and show which noun you are talking about.

direct speech is what is actually spoken by someone. The actual words spoken will be enclosed in **inverted commas**: "Please can I have a drink?"

E

exclamation expresses excitement, emotion or surprise and ends with an exclamation mark.

F

figurative language uses words and ideas to create a mental image. Imagery, metaphors, similes and personification are all types of figurative language.

fronted adverbials are at the start of a sentence. They are usually followed by a comma.

future time is shown in a number of different ways. These all involve the use of a present tense verb.

H

homophones are words that sound the same but are spelled differently and mean different things.

I

imagery uses words that create a picture of ideas in our minds.

inverted commas (also known as speech marks) are punctuation that enclose direct speech: "Please can I have a drink?"

M

main clause: contains a subject and verb and makes sense on its own

metaphors describe something as being something else, even though it is <u>not</u> actually that. The moon was a ghostly galleon.

modal verbs tell us how likely it is that something will happen.

N

nouns are sometimes called 'naming words' because they name people, places and things.

- **proper noun** (Ivan, Wednesday) names something specifically and starts with a capital letter.
- **common noun** (boy, man) names something in general.

noun phrases are phrases with nouns as their main word and may contain adjectives or prepositions: enormous grey elephant/in the garden.

P

parenthesis is a word, clause or phrase inserted into a sentence to add more detail.

past tense describes past events. Most verbs take the suffix ed to form their past tense.

perfect form of a verb usually talks about a past event and uses the verb have + another verb.

- **past perfect**: He had gone to lunch.
- **present perfect**: He has gone to lunch.

personal pronouns replace people or things.

personification is when human qualities are given to an animal, object or thing.

plural means 'more than one'.

possession a word that shows who or what something belongs to using an apostrophe.

possessive pronouns are used to show who something belongs to.

prefix is a set of letters added to the beginning of a word in order to turn it into another word.

prepositions link nouns (or pronouns or noun phrases) to other words in the sentence. Prepositions usually tell you about place, direction or time.

present tense describes actions happening now.

progressive or 'continuous' form of a verb describes events in progress. We are singing.

pronouns are short words used to replace nouns (or noun phrases) so that the noun does not need to be repeated.

- **personal pronouns** replace people or things.
- **possessive pronouns** are used to show who something belongs to.
- **relative pronouns** introduce a relative clause and are used to start a description about a noun.

proper noun (Ivan, Wednesday) names something specifically and starts with a capital letter.

Q

question asks a question, ends with a question mark.

R

relative clause is a type of subordinate clause that changes a noun. It uses relative pronouns such as who, which or that to refer back to that noun.

relative pronouns introduce a relative clause and are used to start a description about a noun.

root word is a word to which new words can be made by adding prefixes and suffixes: happy – unhappy – happiness.

S

sentence is a group of words which have a subject and verb and make sense. There are different types of sentence:

- **statement** is a fact which ends with a full stop.
- **question** asks a question and ends with a question mark.
- **command** tells someone to do something and ends with an exclamation mark or a full stop.
- **exclamation** expresses excitement, emotion or surprise and ends with an exclamation mark.

similes use words such as 'like' or 'as' to make a direct comparison.

singular means 'only one'.

statement is a fact which ends with a full stop.

subordinate clause needs the rest of the sentence to make sense. A subordinate clause includes a conjunction to link it to the main clause.

subordinating conjunctions (when, because) link a subordinate clause to a main clause.

suffix is a word ending or a set of letters added to the end of a word to turn it into another word.

syllable sounds like a beat in a word. Longer words have more than one syllable.

T

tense is **present** or **past** tense and normally shows differences of time.

V

verbs are doing or being words. They describe what is happening in a sentence. Verbs come in different tenses.

vowel sounds are made with the letters a, e, i, o, u. Y can also represent a vowel sound.

W

word families are groups of words that are linked to each other by letter pattern or meaning.

Maths glossary

12-hour clock Time that uses 12 hours, with am before 12 noon, and pm after.

24-hour clock Time that uses 24 hours for the time; does not need am or pm, 5.30pm is written as 17:30.

2D Two-dimensional, a term used for shapes with no depth, usually drawn on paper or viewed on a screen.

3D Three-dimensional, a term used for solid shapes with length, depth and height.

A

Acute angle An angle measuring between 0° and 90°.

Adjacent Near or next to something, usually used for talking about angles, sides or faces.

Analogue clock Shows the time with hands on a dial.

Angle The measure of the gap between lines that meet, or the amount by which something turns; measured in degrees.

Anticlockwise Rotating in the opposite direction to the hands of a clock.

Approximate A number found by rounding or estimating

Area The amount of surface covered by a shape.

Axis (plural *axes*) The horizontal and vertical lines on a graph.

B

Base 10 The structure of our number system; also called powers of 10 because all numbers are based on the powers of 10: 10, 100, 1000, and so on.

C

Clockwise Rotating in the same direction as the hands of a clock.

Coordinates Numbers that give the position of a point on a graph, (x, y).

Cube number A number multiplied by itself twice, such as $2 \times 2 \times 2 = 2^3 = 8$.

D

Decimal fraction A decimal number, whose fraction equivalent has a denominator of a power of 10. For example, 0.5, 0.34, 0.429.

Decimal places Decimals can be written to one or more decimal places: 3.214 has three decimal places.

Decimal point The dot used to separate the fractional part of a number from the whole.

Denominator The number on the bottom of a fraction.

Difference The amount between two numbers.

Digits Our number system uses ten digits, 0–9, to represent all our numbers.

Digital clock A clock that shows time using digits rather than using hands on a dial; can show 12-hour or 24-hour time.

E

Edge The line where two faces of a 3D shape meet.

Equivalent fractions Fractions with different numerators and denominators that represent the same amount, such as $\frac{1}{2}$ and $\frac{2}{4}$.

Estimate To use information to get an approximate answer.

Even numbers Numbers that can be divided by 2; they end in 0, 2, 4, 6 or 8.

Exchange Alternative terminology for carry/borrow used in formal written methods for addition, subtraction, multiplication and division.

F

Face The flat or curved areas of 3D shapes.

Factor A number that will divide exactly into a particular number. For example, 1, 2, 3, 4, 6 and 12 are the factors of 12.

I

Imperial units Units for measuring length, capacity and mass before decimal units were created; length uses inches, feet and miles, capacity uses fluid ounces and pints, and mass uses pounds and tons.

Improper fraction (Also called a vulgar fraction) is a fraction with a numerator larger than its denominator.

Irregular polygon A 2D shape which does not have identical sides and angles.

Isometric drawing A technique for drawing 3D shapes on flat surfaces.

L

Line graphs A graph that shows how something changes over time, like height, temperature or speed.

M

Mental methods Methods for accurately solving calculations without writing them down.

Million The number 1,000,000; one thousand thousand.

Mixed number A whole number and a fraction, such as $3\frac{2}{5}$.

Multiple A number made by multiplying two numbers together, such as: 6 is a multiple of both 2 and 3.

N

Negative number A number less than zero.

Numerator The top number of a fraction; the numerator is divided by the denominator.

O

Obtuse angle An angle measuring more than 90° and less than180°.

Odd numbers Numbers that cannot be divided by 2; they end in 1, 3, 5, 7 or 9.

P

Percentage A number expressed as a fraction out of 100, and represented using the % symbol, meaning per cent. For example, 58%.

Perimeter The distance around the outside edge of a closed shape.

Polygon Any straight-sided 2D shape.

Positive number A number greater than zero.

Powers of 10 The structure of our number system; sometimes called Base ten. Powers of 10 are numbers that are made by multiplying 10 by 10 a number of times. For example, 100 is 10 × 10 or 10 to the power of 2 (10^2), 1000 is 10 × 10 × 10 or 10 to the power of 3 (10^3).

Prime factor A factor that is also a prime number; 3 is a prime factor of 12.

Prime number A whole number that can only be divided by itself and by 1 with no remainder (1 itself is not a prime number).

Proportion The fraction of an amount, such as *eight out of nine people wore red.*

R

Reflection A mirror image of a point or shape, in a mirror line, on a graph.

Reflex angle An angle measuring between 180° and 360°.

Regular polygon A 2D shape with all sides and angles identical.

Roman numerals The system of letters used by the Romans to represent numbers, still sometimes used on clock faces and to represent years.

Rounding Whole numbers are often rounded to the nearest power of ten, and decimals to the nearest whole number, tenth or hundredth to make calculations easier.

S

Square number A number multiplied by itself, such as: $3 \times 3 = 9$ or $3^2 = 9$.

Symbol A sign used for an operation or relationship in mathematics, such as +, −, ×, ÷, =, < and >.

Symmetrical A symmetrical shape is one that is identical either side of a mirror line.

T

Translation Movement of points or shapes on a graph by moving each point by the same amount and in the same direction.

V

Vertex (plural *vertices*) The corner of a 3D shape where edges meet.

English answers

GRAMMATICAL WORDS

Page 10

1 **a.** They could hear the plane's supersonic engine.
b. He filled in the necessary paperwork, before applying for a passport.
c. She had a look of intense concentration on her face.

2 Accept any appropriate adjective which makes sense within sentence, for example:
a. It was a **very delicious** meal.
b. Nikita had **excellent** results in her tests.
c. Their family had a **wonderful** holiday in Majorca.

Page 11

1 **a.** Ellie cautiously opened the dark cupboard.
b. The American group finally reached the top of Mount Everest.

2 Accept any appropriate adjectives, for example:
a. lazy warm afternoon
b. deep cold river
c. fierce terrifying crocodile

Page 12

1 The dog was barking loudly when the postman brought a letter.

2

Present tense	Past tense	Present progressive
he eats	he ate	he is eating
they sleep	they slept	they are sleeping
we run	we ran	we are running

3 They **had finished** their tea when the phone rang.

Page 13

1 Accept either both verbs in the present tense or both verbs in the past tense.

2 **a.** was **b.** are **c.** appears

Page 14

1 **a. May** I go to the bathroom, please?
b. We **could** go to the cinema this afternoon.
c. They **will** be going on holiday on Saturday.

2

Sentence	Modal verb of possibility	Modal verb of certainty
Sunita should tidy her bedroom.	✓	
Sunita must tidy her bedroom.		✓
Sunita might tidy her bedroom.	✓	
Sunita can tidy her bedroom.		✓

Page 15

1 **a.** The waves lapped gently around her feet.
b. The music blared deafeningly from the large speakers.
c. Aggressively, the dog guarded his territory.

2 Accept any appropriate adverb, for example:
a. The snow fell **softly** during the afternoon.
b. The sleeping baby snuffled **quietly**.
c. The footballer **aggressively** tackled the opposing team.

Page 16

1 **a.** (During the morning) she received a telephone call.
b. There was a fire alarm (in the shopping centre.)
c. The cat curled up (very gracefully.)

2

Sentence	Adverbial of time	Adverbial of place	Adverbial of manner
Thomas played the piano really beautifully.			✓
This afternoon we will go to the park.	✓		
The traffic flowed over the bridge.		✓	

Page 17

1 **a.** Opposite sides of a rectangle are **obviously** equal lengths.
b. Maybe the water will be warm enough to swim in.
c. There is **probably** enough petrol in the car.

2 Accept any sentence using an adverb of possibility, appropriately, for example:
We will definitely see you next week.

Page 18

1 During the winter, the geese arrived on the coastal marshes.

2 Accept any sentence with a fronted adverbial, which makes sense, for example: Last night, the winds blew the trees down.

Page 19

1 **a.** Although they read all the information, the committee decided to close all the libraries.
b. There was a flood warning because it had rained a lot.
c. After eating the meal, James was full!

2 **a.** The dragon roared as he blew flames from his mouth.
b. Despite reducing the price, the house didn't sell.
c. School had a cake sale that did very well.

Page 20

1 **a.** We could have a barbecue **or** we could eat inside
b. The carpet is dirty **but** we can clean it.
c. I am playing netball **and** I want to be on the team.

2 Accept any appropriate ending, which is also a subordinate clause, for example:
a. They went to town so **they could buy some shoes**.
b. She stroked the goat yet **she was still scared of it**.
c. I love watching the swans for **they are so graceful**.

Page 21

1 **a.** I wanted to go camping **if** it was sunny.
b. I ran downstairs **because** the doorbell rang.
c. The door creaked open, **then** a hand appeared.

2

Sentence	Co-ordinating conjunction	Subordinating conjunction
I met a school friend **when** I was leaving the library.		✓
We went bowling **and** we went for a pizza.	✓	
You can eat some cake **if** you are hungry.		✓
Would you like to play this game **or** play your new game?	✓	

Page 22

1 **a.** The weather forecast, that we were listening to, told us there would be snow.
b. The man, whose window it was, said it would need to be repaired.
c. The pitch, where the game was to be played, was waterlogged.

2 The sofa

Page 23

1 The hotel

2 **a.** Oscar played with **his** toy engine.
b. I couldn't wait to open **my** presents.
c. The children enjoyed **their** swim.

Page 24

1 **a.** The marathon runner was under a lot of pressure to finish.
b. We had to queue outside the theatre to get tickets.
c. Aisha was between Orla and Gita.

2 Accept any sentence using 'beneath' appropriately, for example: The kitten was hiding beneath the bedclothes.

Page 25

1 **a.** An icy wind blew and many people were hurrying back to their homes.
b. Our accommodation was a disappointment and we telephoned its owner.
c. Jane arranged lots of tables around the garden and waited for her guests to arrive.

2 **a.** I wanted to buy **that** pair of shoes.
b. We need to pack **our** cases.
c. It was her **first** time at gymnastics.

PUNCTUATION

Page 26

1 Where is the nearest petrol station? → Question
I wonder where I will find a petrol station. → Statement

2 Accept any appropriate question starting with 'Who' and ending in a question mark, for example: Who is going to play football today?

3 **a. What** time do we start school?
b. Which is the best way to the beach?
c. When are you going to Scotland?

Page 27

1

Sentence	Statement	Question	Command	Exclamation
Why is the dog barking		✓		
What a beautiful baby				✓
Line up, quietly			✓	
The chocolate ice-cream was delicious	✓			

2 **a.** It was a very exciting game**.**
b. You had an exciting time at Amelia's, didn't you**?**
c. Make it more exciting**!**
d. How exciting**!**

Page 28

1 **a.** we'd **b.** wouldn't **c.** there's

2 **a.** weather is **b.** It is, they will **c.** should have

Page 29

1 **a.** Pippi**'**s food bowl was empty.
b. The children**'**s outing was very successful.
c. The swans**'** care of their cygnets was very touching.

2 The fairies' dresses shimmered in the candles' glow.

Page 30

1 **a.** Someone is being told to tell their cousin called Alex.
b. Someone is being asked if they want to eat Donna!

2 **a.** "Tell your cousin, Alex."
b. "Shall we eat, Donna?"

Page 31

1 **a.** At the end of the street**,** there is a sweet shop.
b. Tomorrow night**,** there will be a full moon.
c. Poorly cooked**,** the food was inedible.

2 Accept any appropriate answer containing a fronted adverbial. For example:
a. Under the tree, you'll find the treasure.
b. In the summer, we will go on holiday.
c. During the storm, the tent blew away.

Page 33

1 **a.** "Today is Monday."
b. "How are you?"
c. "Stop!"
d. "Have you finished your work?" asked the teacher.
e. The teacher asked, "Have you finished your work?"

2 **a.** The teacher looked at the boy and said, "Well done!"
b. "It will rain tomorrow," said the weather forecaster.
c. "Look out!" shouted the driver.
d. "We have some orange juice. We also have some mango juice," said the waiter.

Page 35

1 **a.** Toby **(a six-year-old collie dog)** was lost for seven days.
b. There are many ways**, most of them difficult,** to climb Mount Snowdon.
c. 'Grab a piece of my heart' **– such a great song –** will be number one next week.
d. My new book **– Wheelchair Warrior –** is**, according to my publisher,** going to be a best seller. Accept also: My new book**, Wheelchair Warrior,** is **– according to my publisher –** going to be a best seller.

2 Our favourite place is Venice.

Page 36

1 **a.** Jumila walked slowly towards the door of the house. She did not know what would happen next. She was late and she knew it.
Ten seconds later she was inside facing her father.
b. There is a different place. She is now inside the house.

Page 37

1 We use headings as titles for pieces of writing.
We use subheadings as titles for sections of writing within a longer piece to make information easier to find.

2 Heading: Any appropriate answer that is a summary of the entire passage. For example: Communication Technology
Subheading 1: Any appropriate answer that is a summary of the paragraph. For example: As it was then
Subheading 2: Any appropriate answer that is a summary of the paragraph. For example: How it is now

VOCABULARY

Page 38

1

Prefix	Verb	New verb
dis → appoint	spell	misspell
mis → spell	appoint	disappoint
dis → approve	treat	mistreat
mis → treat	approve	disapprove

2 **a.** misshapen **b.** disembark **c.** mismatch **d.** disbelieve

Page 39

1 **a.** overspend **b.** rearrange **c.** defrost

2 **a.** The spy decoded the message.
b. We reclaimed our baggage after the flight.
c. The car overtook us on the inside lane.

Page 40

1 **a.** originate **b.** medicate **c.** commentate

2 **a.** appreciate **b.** domesticate **c.** demonstrate

Page 41

1 **a.** The butter had started to solidify.
b. The children were able to dramatise the story of Gelert.
c. The farmer needed to fertilise his crops.

2 **a.** individualise **b.** quantify **c.** acidify **d.** terrorise/terrify

3 **a.** terrify/terrorise **b.** popularise **c.** capitalise

Page 43

1

+ prefix	root word	+ suffix
dispossess, repossess	possess	possessed, possesses, possessing, possession
unnatural	natural	naturally, naturalise
misremember	remember	remembering, remembers, remembered, remembrance
disbelieve	believe	believes, believed, believing, believable, believably

2 **a.** accommodate **b.** sincere **c.** solve **d.** continue

3 **a.** appearance **b.** imaginative **c.** comparing **d.** privilege

SPELLING

Page 44

1 **a.**

uff sound	ow (as in cow)	oe (as in toe)	or (as in for)
tough enough	bough	although though	bought fought nought thought

b. trough

2 **a.** I **thought** I would be able to get there in time.
b. The sea was very **rough**.
c. We crawled **through** the tunnel.
d. The boxers **fought** in the ring.
e. **Although** it was very stormy, we managed to reach port.

Page 45

1 **a.** weight **b.** eighth **c.** achieve **d.** neighbour **e.** ceiling

2 **a.** weight **b.** mischievous **c.** neighbour

3 **a.** achieve **b.** thief **c.** perceive **d.** weight **e.** eight
f. retrieve

Page 47

1

Colour each syllable a different colour	What is the tricky bit in this word?
build	i sound: ui
circle	s sound: c at beginning
vehicle	h in the middle: difficult to hear
relevant	e or a in middle ant or ent?
parliament	ia in the middle
environment	n before ment – don't always hear it
restaurant	or sound: au in the middle ant or ent?

2 **a.** calendar **b.** government **c.** twelfth **d.** dictionary
e. business

3 **a.** vegetable **b.** regular **c.** separate **d.** recognise
e. familiar

Page 49

1

One pair of double letters		Two pairs of double letters	More than two pairs of double letters
accident accompany according actually apparent appear appreciate arrive attached business communicate community different difficult disappear equipped grammar occasion opposite possible pressure suppose	correspond especially exaggerate excellent guarantee harass immediately immediate interrupt marvellous necessary occupy occur opportunity profession programme recommend sufficient suggest	accidentally accommodate address aggressive embarrass occasionally possess possession	committee

2 Suggested answers include the following:
a. Opposite
b. impossible
c. occasionally
d. accompany
e. apparent
f. exaggerated

Page 50

1

Root word	Suffix	New word
begin	ing	beginning
forbid	en	forbidden
regret	ed	regretted
limit	ed	limited

Page 51

1 **a.** referring **b.** transferred **c.** referee **d.** preference
e. preferring

Page 52

1 **a.** available **b.** considerable **c.** noticeable **d.** enjoyable

2 **a.** reliably **b.** understandably **c.** comfortably
d. considerably

Page 53

1

	+ ible	+ ibly
force	forcible	forcibly
incredulous	incredible	incredibly
vision	visible	visibly
admission	admissible	admissibly
comprehension	comprehensible	comprehensibly
response	responsible	responsibly

Page 54

1 **a.** yacht **b.** island **c.** doubt **d.** muscle

2 **a.** They rowed the boat towards the deserted isle.
b. I am going to write a story.
c. The lamb was born just after its twin.
d. Dad used the bread knife to cut me a slice.
e. He cut his thumb on the glass.

Page 55

1 **a.** I **practised** the piano every day.
b. The bride walked up the **aisle**.
c. I prepared the **guest** bedroom for the visitors.
d. He walked straight **past** me.
e. **They're** going on holiday next week.

READING

Page 56

1 The main idea is: motor racing is dangerous.

Page 57

1 Dad is not happy with the cat because she has damaged the wallpaper.
Dad is not happy with Mum because she has laughed at the cat's damage.

2 The paragraphs are about Dad getting annoyed so the best answer would be 'Dad's annoyed!'

Page 58

1 **a.** Weekends are wonderful.
b. Any three from the following: no school; no work; nothing to do; 48 hours of selfish laziness; no rush; do what we want.

Page 59

1 **a.** The farmer will use the tractor and the rope to try to pull the car out.
b. The farmer has brought the rope from the tractor.

Page 61

1 The main theme is: bravery/courage.

2 **a.** Possible answers: Pauline set off across the rushing river; threatening to sweep her away; Her torch flickered, fluttered and went out; How would she ever find her way to the treasure now?
b. Possible answers: it creates excitement, it adds drama/excitement to the story, it creates a cliffhanger.
c. The story puts the heroine in a dangerous position and ends with a cliffhanger.

Page 63

1 **a.** Accept feelings such as worried.
b. a worried look on her face.
c. Bad.
d. Accept 'She hid it', or 'She would have to give it to her mother sooner or later but not just yet'.

Page 64

1 raced

2 Accept without success, or similar.

Page 65

1 discourage

2 The distance should put people off going.

Page 67

1 **a.** a simile **b.** steel spears or pools of pain
c.

Example	Type/explanation of figurative language
it hid the teardrops that were swimming from my eyes	This is personification showing how the teardrops moved.
in pools of pain	This is a metaphor showing shape and emotion to describe the size of the pain.
I stumbled home like a blind man	This is a simile showing how difficult it was to see the way.

Page 69

1 **a.** Everyone.
b. Everyone does so I should as well.
c. They make the reader imagine the taste.

2 **a.** chill wind; the graveyard was no place to be in the dark.
b. Makes the reader wonder if they do.
c. Short sentences; asks questions.
d. cold, graveyard, dark, girl alone, strange events.

Page 70

1

Feature	Feature name	Example
Language	Short sentence Repetition of words	Yes? Would you go out
Structural	Heading	Do you think you're brave?
Presentational	Bold	Heading

Page 71

1 As if the writer is talking directly to them.

2 Repeating the word 'dark' increases and emphasises the level of darkness, taking it to the darkest thing imaginable.

Page 73

1 **a.** On the north bank of the Thames.
b. Almost a thousand years ago.
c. Gruesome.
d. The Crown Jewels.
e. Has been a stronghold, has been frightening.

Page 75

1 **a.** Anything relating to comfort.
b. Anything relating to cost or set-up.

2 **a.** Today: hip hop, garage, house
1970s: heavy metal, pop, glam rock
b. Today: Performers are nameless, faceless, personality-less, plastic performers who will be forgotten instantly.
1970s: Real talents. Artists were more memorable. Their music lives on.
c. Record companies will always make the most money.

Page 77

1

	Fact	Opinion
Mount Rushmore National Memorial is one of the most amazing sculptures ever made.		✓
The memorial was carved into the granite face of Mount Rushmore in South Dakota.	✓	
The project ran out of money.	✓	
Such a sculpture will never be achieved again.		✓

2 Any three from the following: Started in 1927, finished in 1941; Four presidents' heads: Washington, Jefferson, Roosevelt and Lincoln; Originally meant to be carved from the waist up.

3 It was a hard choice!

Maths answers

NUMBER AND PLACE VALUE

Page 81

Check

1 thirty-four thousand eight hundred and five

2 237,120

3 forty thousand or four ten thousands

4 7, 12, 725, 25,612, 50,000, 225,421, 899,372, 1,000,000

5 **a.** 3521 < 5630 **b.** 15,204 > 9798
c. 833,521 > 795,732

Brain-teaser Winchcomb City
Brain-buster Winchcomb City, Fintan United, Forest Rovers

Page 83

Check

1 124, 224, 324, 424, 524, 624

2 12,906, 11,906, 10,906, 9906, 8906, 7906

3 320,435, 420,435, 520,435, 620,435, 720,435, 820,435

4 243,000, 233,000, 223,000, 213,000, 203,000, 193,000

Brain-teaser 13 months
Brain-buster 73,456

Page 85

Check

1 **a.** 0 **b.** 0 **c.** –9 **d.** –4

2 –6, –4, –2, 0, 2, 4, 6

3 **a.** – **b.** + **c.** + **d.** –

4 **a.** 4 **b.** 3 **c.** 9 **d.** 2

Brain-teaser –2°C
Brain-buster 36°C

Page 87

Check

	nearest 10	nearest 100	nearest 1000	nearest 10,000	nearest 100,000
67	70	100	0	0	0
145	150	100	0	0	0
3320	3320	3300	3000	0	0
78,249	78,250	78,200	78,000	80,000	100,000
381,082	381,080	381,100	381,000	380,000	400,000
555,555	555,560	555,600	556,000	560,000	600,000

Brain-teaser 50,000
Brain-buster Loss on a bad match: £46,000; Gain on a good match: £54,000

Page 88

Check

1 **a.** 8 **b.** 23 **c.** 300 **d.** 95 **e.** 104 **f.** 140 **g.** 610
h. 900

2 **a.** XXII **b.** XLI **c.** LV **d.** XCIII **e.** CXII **f.** CLX
g. CCXII **h.** CMLXV

Brainbuster Date: CDX Years in Britain: CDLXV

CALCULATIONS

Page 89

Check

1 **a.** 96 **b.** 226 **c.** 5276 **d.** 9954 **e.** 130,320

2 **a.** 34 **b.** 95 **c.** 246 **d.** 1500 **e.** 265,675

Brain-teaser 273
Brain-buster £1097

Page 91

Check

1 **a.** 3244 **b.** 12,309 **c.** 70,180 **d.** 621,229

2 **a.** 5966 **b.** 69,636 **c.** 213,925 **d.** 658,930

Brain-teaser Yes
Brain-buster No

Page 93

Check

1 **a.** 139 **b.** 4163 **c.** 189,419

2 **a.** 119 **b.** 2273 **c.** 3968 **d.** 3861 **e.** 116,923

Brain-teaser Bim and Bom
Brain-buster 212,980

Page 95

Check

1 1, 2, 3, 6

2 4, 8, 12, 16, 20, and so on.

3 **a.** 1 × 15 and 3 × 5 **b.** 1 × 27 and 3 × 9
c. 1 × 24, 2 × 12, 3 × 8 and 4 × 6
d. 1 × 30, 2 × 15, 3 × 10 and 5 x 6.

4 **a.** 1, 2 and 4 **b.** 1 and 5 **c.** 1, 2 and 4
d. 1, 2, 5, 10, 25 and 50

Brain-teaser

Children	1	2	3	4	6	8	12	24
Chocolates	24	12	8	6	4	3	2	1

Shared between 5 children there would be 4 chocolates left over.
Brain-buster No, because 7 is not a factor of 365,
52 × 7 = 364

Page 97

Check

1 A number that can only be divided by itself and 1.

2 2, 3, 5, 7, 11, 13, 17, 19

3 **a.** 25 is not a prime because 5 is a factor **b.** 71 is a prime because it can only be divided by itself and 1 **c.** 87 is not a prime because 3 and 29 are factors

4 Many possible answers, such as 101, 103, 107, 109, 113

Brain-teaser No, because it can be divided by 7 and 11.
Brain-buster 27 is not a prime number because it can be divided by 3 and 9.

Page 99

Check

1 **a.** 273 **b.** 552 **c.** 1395 **d.** 3528

2 **a.** 315 **b.** 832 **c.** 795 **d.** 1320

Brain-teaser £6.45
Brain-buster Yes, because 475 × £13 = £6175

Page 101

Check

1 **a.** 19 **b.** 11r2 **c.** 46r4 **d.** 32r2

2 **a.** 14 **b.** 25 **c.** 65r2 **d.** 21r3

Brain-teaser 12 times
Brain-buster 11 pieces per child. Teacher has 15 pieces.

Page 102

Check

1 **a.** 200 **b.** 960 **c.** 8888 **d.** 30,000 **e.** 14,350

2 **a.** 50 **b.** 43 **c.** 1001 **d.** 902 **e.** 404

Brain-buster £2050 each

Page 103

Check

1	2	3	4	5	6	7	8	9	10
1^2	2^2	3^2	4^2	5^2	6^2	7^2	8^2	9^2	10^2
1×1	2×2	3×3	4×4	5×5	6×6	7×7	8×8	9×9	10×10
1	4	9	16	25	36	49	64	81	100
1^3	2^3	3^3	4^3	5^3	6^3	7^3	8^3	9^3	10^3
1×1×1	2×2×2	3×3×3	4×4×4	5×5×5	6×6×6	7×7×7	8×8×8	9×9×9	10×10×10
1	8	27	64	125	216	343	512	729	1000

Brain-teaser 25 goals
Brain-buster 9 × 9 × 9 = 729 apples

Page 105

Check

		×10	×100	×1000
	3	30	300	3000
÷10	0.3	3	30	300
÷100	0.03	0.3	3	30
÷1000	0.003	0.03	0.3	3

		×10	×100	×1000
	27	270	2700	27000
÷10	2.7	27	270	2700
÷100	0.27	2.7	27	270
÷1000	0.027	0.27	2.7	27

		×10	×100	×1000
	48	480	4800	48000
÷10	4.8	48	480	4800
÷100	0.48	4.8	48	480
÷1000	0.048	0.48	4.8	48

		×10	×100	×1000
	317	3170	31700	317000
÷10	31.7	317	3170	31700
÷100	3.17	31.7	317	3170
÷1000	0.317	3.17	31.7	317

Brain-teaser 32,000 feet
Brain-buster 1.356kg

Page 107

Check

1 **a.** 3 cakes **b.** 5 adults **c.** 22 animals **d.** 75 children

2 Children's drawing of:
a rectangle 3 squares wide and 2 deep
a rectangle 5 squares wide and 1 and a half squares deep

3

Item	Room	Table	Chair	Cupboard	Basket
Real height	280cm	90cm	40cm	170cm	25cm
Model height	14cm	4.5cm	2cm	8.5cm	1.25cm

Brain-teaser 3600 beats in an hour; 86,400 beats in a day
Brain-buster Model scale: $\frac{1}{15}$

Page 109

Check

1 **a.** 9 **b.** 12 **c.** 6 **d.** 0

2 **a.** correct **b.** wrong (8) **c.** correct **d.** wrong (9) **e.** correct **f.** wrong (5)

3 **a.** ÷ **b.** ÷ and –

Brain-teaser £3.10
Brain-buster £4.30

FRACTIONS, DECIMALS AND PERCENTAGES

Page 111

Check

1 **a.** $\frac{4}{8}$ **b.** $\frac{2}{8}$ **c.** $\frac{6}{8}$ **d.** $\frac{8}{8}$

2 **a.** $\frac{6}{12}$ **b.** $\frac{3}{12}$ **c.** $\frac{8}{12}$ **d.** $\frac{10}{12}$

3 **a.** True **b.** False **c.** True **d.** True **e.** True **f.** False **g.** True **h.** False **i.** False **j.** True **k.** True **l.** False

4 **a.** $\frac{1}{10}, \frac{1}{6}, \frac{1}{5}, \frac{1}{4}, \frac{1}{3}, \frac{1}{2}$ **b.** $\frac{3}{5}, \frac{5}{8}, \frac{3}{4}$ **c.** $\frac{4}{7}, \frac{2}{3}, \frac{7}{9}$

Brain-teaser $\frac{2}{5}$
Brain-buster neutral ($\frac{5}{21}$) < red ($\frac{7}{21}$) < blue ($\frac{9}{21}$) **or** neutral ($\frac{5}{21}$) < red ($\frac{1}{3}$) < blue ($\frac{3}{7}$)

Page 113

Check

1 **a.** $\frac{7}{2}$ **b.** $\frac{9}{4}$ **c.** $\frac{21}{5}$ **d.** $\frac{4}{3}$ **e.** $\frac{8}{3}$ **f.** $\frac{11}{4}$ **g.** $\frac{19}{5}$ **h.** $\frac{17}{2}$

2 **a.** $1\frac{1}{2}$ **b.** $1\frac{1}{3}$ **c.** $1\frac{1}{4}$ **d.** $1\frac{1}{5}$ **e.** $3\frac{2}{3}$ **f.** $1\frac{3}{4}$ **g.** $7\frac{1}{2}$ **h.** $2\frac{3}{5}$

3 **a.** $\frac{3}{2} < 2\frac{1}{2}$ **b.** $\frac{4}{3} = 1\frac{1}{3}$ **c.** $\frac{7}{4} > 1\frac{1}{4}$ **d.** $\frac{13}{2} < 7\frac{1}{2}$
e. $6\frac{1}{4} = \frac{25}{4}$ **f.** $3\frac{1}{2} < \frac{8}{2}$ **g.** $\frac{10}{3} > 2\frac{2}{3}$ **h.** $3\frac{1}{5} > \frac{12}{5}$

Brain-teaser Ali
Brain-buster $\frac{3}{4}$ a pizza

Page 115

Check

1 **a.** $\frac{3}{5}$ **b.** $\frac{5}{7}$ **c.** $\frac{4}{4}$ or 1 **d.** $1\frac{1}{5}$

2 **a.** $\frac{1}{6}$ **b.** $\frac{3}{8}$ **c.** $\frac{2}{4} = \frac{1}{2}$ **d.** $\frac{6}{20} = \frac{3}{10}$

3 **a.** $\frac{3}{4}$ **b.** $\frac{5}{6}$ **c.** $\frac{3}{6} = \frac{1}{2}$ **d.** $\frac{29}{30}$

4 **a.** $\frac{1}{4}$ **b.** $\frac{1}{6}$ **c.** $\frac{1}{6}$ **d.** $\frac{19}{30}$

Brain-teaser $\frac{3}{6}$ or $\frac{1}{2}$ a pizza
Brain-buster $\frac{5}{12}$

Page 117

Check

1 **a.** 5 **b.** 2 **c.** 6 **d.** 10

2 **a.** $2\frac{1}{2}$ or $2\frac{2}{4}$ **b.** $1\frac{1}{2}$ **c.** $1\frac{5}{7}$ **d.** $3\frac{1}{3}$

3 **a.** 7 **b.** $4\frac{1}{2}$ **c.** $10\frac{5}{6}$ **d.** $26\frac{2}{3}$

Brain-teaser $19\frac{1}{2}$ logs
Brain-buster 8 laps

Page 119

Check

1 **a.**

Fraction	$\frac{1}{10}$	$\frac{2}{10}$	$\frac{3}{10}$	$\frac{4}{10}$	$\frac{5}{10}$	$\frac{6}{10}$	$\frac{7}{10}$	$\frac{8}{10}$	$\frac{9}{10}$	$\frac{10}{10}$
Decimal	0.1	0.2	0.3	0.4	0.5	0.6	0.7	0.8	0.9	1.0

b.

Fraction	$\frac{1}{5}$	$\frac{2}{5}$	$\frac{3}{5}$	$\frac{4}{5}$	$\frac{5}{5}$
Decimal	0.2	0.4	0.6	0.8	1

2 **a.** 0.5 **b.** 0.75 **c.** 0.1

3 **a.** $\frac{1}{4}$ **b.** $\frac{7}{10}$ **c.** $\frac{4}{10}$

Brain-teaser $\frac{3}{4}$
Brain-buster No, one quarter = 0.25

Page 121

Check

1

Fraction name	Decimal Fraction	Decimal	Decimal name
five tenths	$\frac{5}{10}$	0.5	zero point five
twenty-three hundredths	$\frac{23}{100}$	0.23	zero point two three
four hundred and thirty-five thousandths	$\frac{435}{1000}$	0.435	zero point four three five
Three tenths	$\frac{3}{10}$	0.3	zero point three
Eighty-six hundredths	$\frac{86}{100}$	0.86	zero point eight six
Five hundred and seven thousandths	$\frac{507}{1000}$	0.507	zero point five zero seven
Eight tenths	$\frac{8}{10}$	0.8	zero point eight
One hundred and thirty-two thousandths	$\frac{132}{1000}$	0.132	zero point one three two
Thirty-nine hundredths	$\frac{39}{100}$	0.39	zero point three nine
One hundred and four thousandths	$\frac{104}{1000}$	0.104	zero point one zero four

Brain-teaser $\frac{87}{100}$, 0.204

Brain-buster $\frac{765}{1000}$, 0.765

Page 123

Check

1 **a.** 0.465 **b.** 0.204

2 **a.** 0.6 **b.** 0.12 **c.** 0.325

3 **a.** zero point three nine five **b.** zero point six zero two **c.** zero point zero zero five

4 **a.** 0.002, 0.084, 0.146, 0.327, 0.5, 0.75, 0.807, 0.999
b. Check numbers on number line are accurately positioned.

Brain-teaser Bugs B and C
Brain-buster Bug C: 0.009cm, Bug B: 0.029cm, Bug D: 0.031cm, Bug A: 0.101cm

Page 125

Check

1 **a.** 1 **b.** 2 **c.** 5 **d.** 0 **e.** 1 **f.** 7 **g.** 12 **h.** 7

2 **a.** 0.8 **b.** 0.8 **c.** 0.5 **d.** 0.8 **e.** 5.6 **f.** 4.1 **g.** 12.8 **h.** 7.0

3 Rounding decimals makes it easier to calculate amounts, but the answers will not be accurate.

Brain-teaser

Day	Monday	Tuesday	Wednesday	Thursday	Friday	Saturday
Money	£52.14	£45.61	£60.13	£46.50	£72.24	£35.51
Rounded	£52	£46	£60	£47	£72	£36

Brain-buster More, because he rounds up more than he rounds down (rounded to £313; exact amount £312.13).

Page 127

Check

10%	0.37	$\frac{1}{2}$
25%	1.0	$\frac{1}{10}$
30%	0.1	$\frac{1}{4}$
37%	0.3	$\frac{100}{100}$
50%	0.75	$\frac{3}{10}$
60%	0.25	$\frac{3}{5}$
75%	0.5	$\frac{37}{10}$
100%	0.6	$\frac{3}{4}$

Brain-teaser 15 have dinners, 10% have sandwiches
Brain-buster 12 children go home for lunch. This is 40% of the class.

MEASUREMENT

Page 129

Check

1 **a.**

cm	m
10	1
25	2.5
52	5.2
100	10
300	30
170	17
60	6
2	0.2

b.

mm	cm
100	1
35	0.35
450	4.5
1000	10
8000	80
900	9
90	0.9
27	0.27

c.

m	km
1000	1
250	0.25
5350	5.35
10,000	10
6000	6
4500	4.5
1350	1.35
4	0.004

2

Imperial	Metric
1 inch	2.54cm
5 inches	12.7cm

Imperial	Metric
10 inches	25.4cm
1 mile	1.61km

Imperial	metric
100 miles	161km
300 miles	483km

Brain-teaser 4.8m
Brain-buster 5.35km

Page 131

Check

1 **a.** 10cm **b.** 8cm **c.** 12cm

2 **a.** 24cm **b.** 24cm

3

shape	formula	length	width	perimeter
rectangle	p = 2(l+ w)	6mm	3mm	18mm
square	p = 4s	2.5mm	2.5mm	10m (1cm)

4 **a.** w = 1cm **b.** w = 8m

Brain-teaser 3m
Brain-buster 7m

Page 133

Check

1 **a.** approx 15cm^2 **b.** approx 8cm^2 **c.** approx 13cm^2

2 **a.** 108cm^2 **b.** 49m^2 **c.** 300m^2

3 **a.** a rectangle, length 5cm and width 1cm **b.** a square of side 5cm **c.** a square of side 6m **d.** a rectangle, length 17km and width 9km

Brain-teaser 14m^2
Brain-buster 81,000 potatoes

Page 135

Check

1

object	grams	kilograms
child	50,000g	50kg
dog	12,000g	12kg
pencil	75g	0.075kg
book	408g	0.408kg

2

object	millilitres	litres
teapot	1250ml	1.25l
sink	8500ml	8.5l
mug	125ml	0.125l
thimble	12ml	0.012l

3 **a.** 483g or 0.483kg **b.** 137ml or 0.137l **c.** 50.408kg or 50,408g **d.** 1.262l or 1262ml

Brain-teaser 20kg
Brain-buster 851ml or 0.851l

Page 137

Check

12-hour	midnight	2:10am	9:15am	noon	3:30pm	9pm	11:59pm
24-hour	00:00	02:10	09:15	12:00	15:30	21:00	23:59

2 **a.** 180 minutes **b.** 48 hours **c.** 84 days **d.** 1460 days (or +1 if leap year included) **e.** 1 hour 40 minutes **f.** 4 days 4 hours **g.** 3 hours 45 minutes **h.** 1 year 135 days

3 **a.** 11 hours and 55 minutes **b.** 4 days 12 hours and 41 minutes **c.** 9 days 2 hours and 58 minutes

Brain-teaser 8:37am
Brain-buster 604,800 seconds

Page 139

Check

1

Pence	200p	135p	6325p	9p	10,903p
Pounds	£2	£1.35	£63.25	£0.09	£109.03

2

Pounds	£4	£2.56	£0.12	£82	£403.20
Pence	400p	256p	12p	8200p	40,320p

3 **a.** £7.77 **b.** £22.15 **c.** £1.70 **d.** £41.73 **e.** £6.70 **f.** £70.40 **g.** £12.50 **h.** £7.70

Brain-teaser £15.75
Brain-buster £600

GEOMETRY

Page 141

Check

1 **a.** 90° right-angle **b.** 130° obtuse angle **c.** 65° acute angle

2 **a.** right-angle **b.** reflex angle **c.** acute angle

Brain-teaser Aaron can place his ruler alongside the line to decide if it is more or less than 180°. Over 180° is reflex, less than 180° is acute or obtuse. If he rests his pencil at right angles to his ruler this will give him a right angle. Angles less than this will be acute; more than this but less than 180° will be obtuse.
Brain-buster Because the angles on a straight line add up to 180°.

Page 143

Check

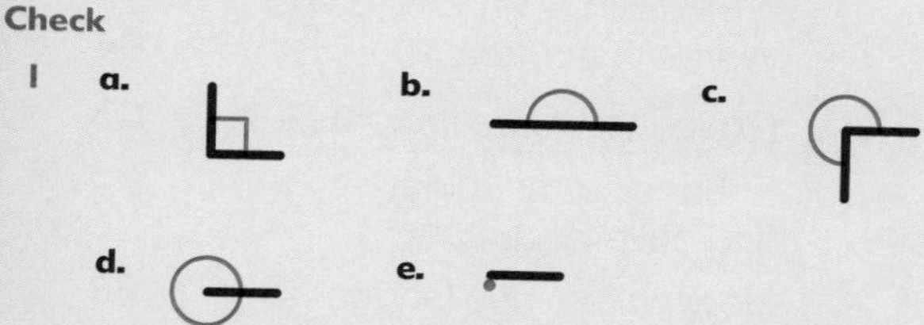

Brain-teaser 3, 270°
Brain-buster 180° anticlockwise

Page 145

Check

1 **a.** regular triangle **b.** regular hexagon **c.** irregular quadrilateral **d.** irregular pentagon

2 Because the sides are not all the same length.

3 **a.** square **b.** parallelogram **c.** kite **d.** rectangle **e.** trapezium **f.** rhombus

Brain-teaser Both long sides should be labelled 6cm, both short sides should be labelled 4cm. All four angles should be labelled 90°.

Page 147

Check

1

triangular prism	cylinder	cube	cuboid

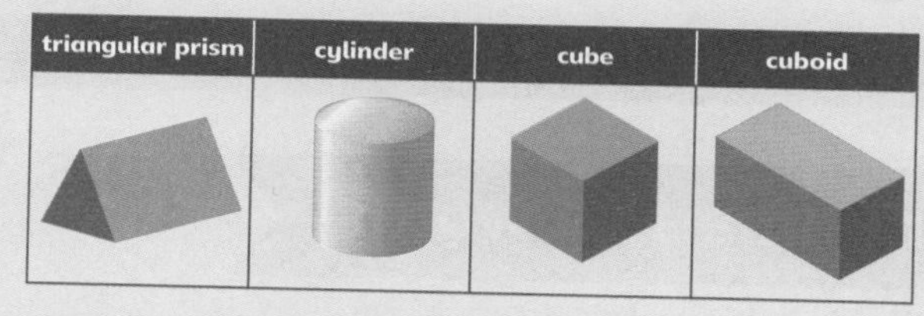

2 **a.** sphere **b.** cube **c.** cone **d.** cylinder **e.** triangular prism **f.** cuboid

Brain-teaser cuboid
Brain-buster The cylinder would hold more water than the cone because it is the same width all the way along, but the cone gets narrower.

Page 149

Check

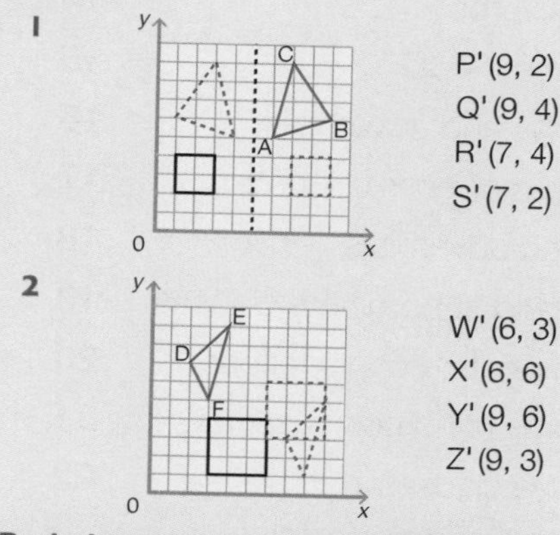

1 P' (9, 2)
Q' (9, 4)
R' (7, 4)
S' (7, 2)

2 W' (6, 3)
X' (6, 6)
Y' (9, 6)
Z' (9, 3)

Brain-buster If the mirror line is horizontal the point moves up or down, so the *y*-coordinate changes. If it is vertical they move left or right, so the *x*-coordinate changes. Find the distance the point is from the mirror line, then move the x or y coordinate by double the amount.

STATISTICS

Page 151

Check

1 **a.**

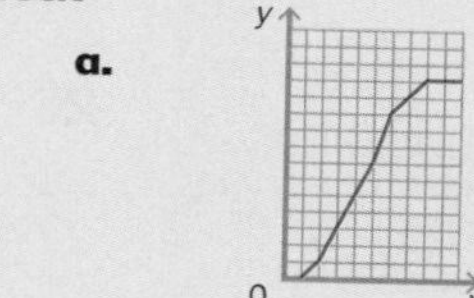

b. 3 weeks **c.** 8 weeks **d.** 6 weeks

Brain-teaser Highest temperature: 18°C
When was the temperature highest? 8am
Brain-buster Find the difference between the highest and lowest temperatures: 16°C

Page 153

1 **a.** Earth **b.** Mars **c.** Mercury **d.** Earth and Mars

2 **a.** number 7 bus **b.** every half hour **c.** 3 minutes **d.** The train station could be further away than the supermarket, or the number 6 bus may have to wait at the train station for a set amount of time.

Brain-teaser The number 7 at 10:15
Brain-buster 14:45

English progress tracker

Grammatical words

Punctuation

Vocabulary

Spelling

Reading